Starch in the Bioeconomy

Starch in the Bioeconomy

Jean-Luc Wertz and Bénédicte Goffin

CRC Press
Taylor & Francis Group
Boca Raton London New York

CRC Press is an imprint of the
Taylor & Francis Group, an **informa** business

First edition published 2021
by CRC Press
6000 Broken Sound Parkway NW, Suite 300, Boca Raton, FL 33487-2742

and by CRC Press
2 Park Square, Milton Park, Abingdon, Oxon, OX14 4RN

© 2021 Taylor & Francis Group, LLC

First edition published by CRC Press 2021

CRC Press is an imprint of Taylor & Francis Group, LLC

Library of Congress Cataloging-in-Publication Data
Names: Wertz, Jean-Luc, author. | Goffin, Bénédicte, author.
Title: Starch in the bioeconomy / by Jean-Luc Wertz and Bénédicte Goffin.
Description: First edition. | Boca Raton, FL : CRC Press/Taylor & Francis
Group, LLC, 2020. | Includes bibliographical references and index. |
Summary: "Starch is the most widespread and abundant reserve
carbohydrate in plants and is unique in that it can be used for food,
materials in biobased products, and energy. This book covers structure,
biosynthesis, biodegradation, properties, and applications of starch in
the context of the bioeconomy. The book is aimed at researchers and
industry professionals focused on the development of starch science,
technology, and economics. Built on a reliable and well-documented base
of information, the book presents the paths that remain to be taken to
decipher this still mysterious resource that has contributed so much to
the rise of humanity"—Provided by publisher.
Identifiers: LCCN 2020043489 (print) | LCCN 2020043490 (ebook) |
ISBN 9780367630409 (hardback) | ISBN 9780367630416 (pbk) |
ISBN 9781003111986 (ebook)
Subjects: LCSH: Starch. | Starch industry.
Classification: LCC TP248.S7 W47 2020 (print) | LCC TP248.S7 (ebook) |
DDC 664/.2—dc23
LC record available at https://lccn.loc.gov/2020043489
LC ebook record available at https://lccn.loc.gov/2020043490

ISBN: 978-0-367-63040-9 (hbk)
ISBN: 978-1-003-11198-6 (ebk)

Typeset in Times
by codeMantra

Visit the eResources: https://www.routledge.com/Starch-in-the-Bioeconomy/Wertz-Goffin/p/book/9780367630409

Contents

Preface

Over the past two centuries, there have been monumental advances in nutrition, allowing the virtual elimination of famine in the developed world. However, these 200 years are just a small episode in the long co-evolution between humans and plants over several millions of years. Within this relationship, the most significant advances took place about 15,000 years ago, when plant domestication took place in various regions around the globe.

Rice is unique among wild plants for having been domesticated independently on three continents, Asia, Africa, and South America, as researchers have recently discovered. Asia is believed to be the home of domesticated rice. The oldest evidence of rice consumption identified to date is four grains of rice recovered from the Yuchanyan Cave, Hunan Province in China, dated to between 12,000 and 16,000 years ago. It is generally agreed that the original domestication for all varieties of rice took place in the lower Yangtze River Valley by hunter-gatherers approximately 9,000 to 10,000 years ago. In Africa, domestication/hybridization happened about 3,200 years ago, during the African Iron Age in the Niger Delta region of West Africa. The South American variety was cultivated about 4,000 years ago, in the southwestern Amazon basin; however, it was abandoned after the Europeans arrived. The Americas are considered to be the home of the domestication of maize and potatoes (*Solanum tuberosum*). In Meso-America, the earliest dated maize cob was discovered in Guilá Naquitz cave in Oaxaca dating back to 4300 BC. It was first domesticated in South America, in the Andean highlands, between Peru and Bolivia, more than 10,000 years ago. Wheat and barley (*Hordeum vulgare*) are two of the founder crops of the agricultural revolution that took place 10,000 years ago in the Fertile Crescent, and both remain among the world's most important crops.

The ability to domesticate and cultivate plants close to where humans lived radically changed our way of living. As food became more reliable, permanent settlements were established, and populations grew. In their quest to improve yields and secure crop production, early farmers continually selected the best plants to grow. They continued this process generation after generation until the nature of the plant was modified to a point where many species became unable to survive and compete in the world. While these domesticated plants became dependent on humans, the human also had to adapt biochemically. Our genes evolved and our metabolism also changed to match our diet and lifestyle.

The effect of the dietary changes resulting from an increase in starch consumption offers the best-documented example of the co-evolution of plants and humans in terms of the human-specific adaptation. As humans went from being hunter-gatherers to become farmers, it became necessary to enhance ways to extract as much nutrition from the starch that now comprised a significant proportion of the typical diet. The digestive process of breaking down starch starts as soon as it is ingested, under the action of the salivary amylase. This enzyme specifically breaks down starch into simpler sugars that can be processed by other enzymes or absorbed by the intestine.

Salivary amylase comprises a small fraction of the total amylase secreted by the body, the remainder is mostly made by the pancreas. However, native starch shows high resistance to enzymatic hydrolysis and needs to be cooked before its consumption, to help disrupt the components of the starch granule, because amylase cannot degrade the double-helical starch structure.

The amylase genes (*AMY*), which encode for a starch-digesting enzyme in humans, went through several evolutions. Humans have multiple copies of the gene, but the exact number of copies varies from person to person. Some people have two, others twenty, with an average of six to eight. Researchers have found that groups whose ancestors tended to eat more starch have more salivary amylase, that is, the descendants of farmers rather than hunter-gatherers of the desert.

In contrast, hunter-gatherers of the rainforest or people of the cold north, who mostly relied on fish and fats, do not tend to produce much salivary amylase. Populations with little starch in their diets also have a relatively low copy number of amylase genes (five per individual). Those eating starch food have a higher number with up to seven amylase genes per individual. The difference is small, and the distributions also overlap significantly; however, the difference is measurable and significant. It implies that there may have been some selection for higher copy numbers in cultures with diets rich in starchy plants.

This example of ongoing human evolution raises the question of how this difference might have come about. Individuals with higher copy numbers of the gene might have had an evolutionary advantage because of being able to more effectively digest starch. People in this group who could deal better with starchy foods were more successful. Success is associated with the capability to get more nutrition from their food, for example, a higher level of glucose in the mother's blood during pregnancy or a higher chance of survival during hard times. Eating starch did not cause more amylase genes to appear. People producing more salivary amylase simply did better and had more kids. Over time, the population shifted to having more people with extra amylase genes which is typical of natural selection at work.

The history of potatoes throughout their domestication, export, introduction, and contribution to contemporary society illustrates the uniqueness of starch in the production of food, materials, and energy, as a unique renewable biological resource. Today, the potato is the fifth most important crop worldwide, after wheat, corn, rice, and sugar cane.

Potatoes were first domesticated in the Andean highlands as early as 13,000 years before the present day. Potatoes spread slowly to the coasts and throughout the rest of America before reaching Mexico about 6,000 to 5,000 years ago, probably passing through Lower Central America and the Caribbean Islands. Following the discovery of the Americas in 1492, the potato was then soon introduced to Europe by the Spanish in the late 16th century. The first cultivations took place in the Spanish Canary Islands around the 1570s. Subsequently, potatoes were cultivated by Spanish farmers and exported to France and the Netherlands. The first scientific description of the potato appeared in 1596 when the Swiss naturalist Gaspard Bauhin awarded it the name *Solanum tuberosum esculentum* (later simplified to *Solanum tuberosum*).

The spread of the potato throughout Europe was rapid, by historical standards. In England, the first record of potatoes being cultivated as a staple crop was in the 1690s. Adoption was probably encouraged by failures of existing crops during the "Little Ice Age" and the wars and famines of the 18th century. There was, however, some reluctance to its adoption with the potato generally viewed as a poisonous and dirty plant, or as a strange gift and botanical curiosity. For example, when famine hit Prussia in 1744, King Frederick II attempted to engage his subjects with potatoes. Receiving a less than rapturous response, he used reverse psychology to make potatoes seem "valuable enough to steal." He ordered the plantation of the tuber in a particular field in a secretive fashion. By positioning his guards around it, he made the people think there was something desirable in that royal soil. Some opportunists took the chance to rummage in the dirt, from where they pulled out potatoes before realizing the versatility of potato and its levels of resilience. Nevertheless, it proved a masterstroke for European population. While serving in the French army, during the Seven Years' War, Antoine Parmentier was captured and imprisoned by the Prussians. In prison, he ate little but potatoes, and this diet kept him in good health. Trained as a pharmacist, he devoted the rest of his life to the promotion of *Solanum tuberosum*. At the end of the 18th century, France was hit by high price controls on cereal which caused several hundred civil disturbances. Parmentier proclaimed the nation would stop fighting over bread if its citizens would only eat potatoes. He succeeded in promoting the idea of planting vast areas with clones—a true monoculture, and for the first time in European history, a definitive solution was found to end famine. Compared with grains, potato tubers are inherently more productive. If the head of a wheat or rice plant grows too big, the plant will fall over, with fatal results. With potatoes, the rest of the plant does not limit the underground growth of tuber. Potatoes were so productive; the practical effect, in terms of calories, was to double the food supply of Europe. The resulting increase in calorific intake after the middle of the 18th century had a major influence on the reduction in mortality and a subsequent effect on population growth, as well as on urbanization. The main beneficiaries of the introduction of potatoes were primarily Eastern European countries. In the article 'How the Potato Changed World History', it is argued that "potatoes, by feeding rapidly growing populations, permitted a handful of European nations to assert domination over most of the world between 1750 and 1950."

However, the lack of genetic diversity due to the low number of varieties left the crop vulnerable to disease. A strain of potato blight (*Phytophthora infestans*) originated from Peru spread to Europe in the 1840s. It destroyed potatoes in the Netherlands, Germany, Denmark, and England. In Ireland which had adopted the tuber as its national dish, *P. infestans* wiped out the equivalent of one-half to three-quarters of a million acres that were planted in 1845. The next year was even worse, as was the year after that. The attack did not wind down until 1852. The effect of the Great Famine was devastating, causing one million deaths, up to two million refugees, spurring a century-long population decline. Another infection came from an imported species *Leptinotarsa decemlineata*, the orange-and-black potato beetle, which encountered the cultivated potato around the Missouri River in the early 1860s. Because growers planted just a few varieties, pests such as the beetle and diseases like the blight had a narrower range of natural defenses to overcome, a task even made easier by the developments of railroads and steamships which helped to spread pests and diseases. There were the many desperate trials by farmers to get rid of the bug, but serendipitously one threw some leftover green paint (so-called Paris green) on his infected potatoes. The arsenic and copper contained in the pigments proved to be effective in killing the bug, to the relief of farmers, who then sprayed their potatoes with dilute Paris-green solutions. The success attracted the attention of the chemistry community, and candidate treatments for the control of blight emerged in 1882 with the discovery of the fungicidal spray of copper sulfate and quicklime (Bordeaux mixture). The first spray trials with the Bordeaux mixture for the control of potato late blight occurred in 1886 and the pesticide industry was born!

Although the early history of starch usage is largely unrecorded, some very early examples of its industrial use are documented. Egyptians in the pre-dynastic period (4000 BC) cemented strips of papyrus together with starch adhesive made from wheat. Later, Chinese documents were first coated with a high fluidity starch to provide resistance to ink penetration, then covered with powder starch to give weight and thickness. In the Middle Ages, starch found its principal use in the laundry for stiffening fabrics. The custom of powdering the hair with starch appears to have become popular in France in the 16th century, and by the end of the 18th century, this became a general practice.

However, during these times, none of these applications of starch as a material have jeopardized its use as a nutritional element. However, a recent discovery highlighted the competition between the use of rice as a material and its use for nutritional purposes, in the world's first known example of a composite mortar. Under the Ming dynasty, workers built sections of the Great Wall about 600 years ago by mixing a paste of sticky rice flour and slaked lime, the standard ingredient in mortar. This resulting ancient mortar was a special kind of organic and inorganic mixture: the inorganic component, calcium carbonate, and the organic part, amylopectin, which came from the sticky rice soup added to the mortar. Amylopectin helped create a compact microstructure, giving more stable physical properties and a higher mechanical strength. The sticky rice mortar has bound the bricks together so tightly that in many places, weeds still cannot grow. However, this technical innovation was met with widespread resentment against the Wall in the south of China, because of the Ming emperor having requisitioned the southern rice harvest, both to feed the workers on and to make the mortar. The use of sticky rice, a staple in East Asian food, was one of the highest technical innovations of the time and helped the Ming dynasty build tombs, pagodas, and walls, and weather earthquakes and other disasters.

Starch has been used over several millennia for many different applications. However, research on understanding this substance only spans about three centuries starting with Leeuwenhoek who observed it microscopically as discrete granules in 1716.

About 250 million years ago, the world consisted of a single giant landmass now known as Pangaea. Geological forces broke Pangaea apart, creating the continents and hemispheres familiar today. Over the four eons, the separate corners of the earth have developed wildly different repertoires of plants and animals. Nevertheless, evolution retained the starch granule as Nature's main way to store energy in green plants over long periods. The granule is well suited to this role, being insoluble in water, and densely packed, but still available to the plant's catabolic enzymes.

Starch granules are mainly found in seeds, roots, and tubers, as well as in stems, leaves, fruits, and even pollen. The granules occur in all shapes and sizes (spheres, ellipsoids, polygons, platelets, and irregular tubules); they have diameters ranging from around 0.1 to 200 μm, depending on the botanical source. The differences in granule external morphologies are generally sufficient to allow unambiguous characterization of the botanical source, via optical microscopy (Figure 1).

Irrespective of their botanical origin and their diversity, it is remarkable to find that the internal structures of starch granules share universal features. Native starch granules exhibit a Maltese cross when observed in polarised light. Such a feature points toward the existence of a radial organization of crystalline-type arrangements, having dimensions causing optical polarization, because the visible optical polarisation is in the order of the wavelength of visible light (100 to 1000 nm). The results from several decades of intense investigation have uncovered that the highly complex, hierarchical structure of the starch granule, over six orders of magnitudes, slowly unravels. Among the best-established structural elements is the double-helical structure which is formed by a parallel arrangement of single left-handed strands. Various pieces of fundamental information gathered from experimental characterization can be tentatively connected to help explain the architecture of the starch granule. In doing so, an overall picture can be drawn that shows a dramatic similarity between two processes fundamental for life: storing and transmitting information on the one hand and energy storage on the other.

DNA's ability to store and transmit information lies in the fact that it consists of two antiparallel polynucleotide strands that twist around each other to form a double-stranded helix (Figure 2). The universality of information machinery extends from the double helix to the chromosome from 2 nm to 1200 nm. The ability of starch to store energy in plants lies in the organization of its complex architecture, with several well-organized strata that maintain the compacity over structures that develop on six orders of magnitudes. The universal architecture found in plants extends from the double helix to the starch granule, i.e. from 2 nm to 0.1 to 200 μm. While being structurally robust, this organization is flexible enough to embrace a multitude of variations that express the diversity of the plant kingdom and is the basis of the many functional properties of the starch granule.

Starch remains more than ever vital for our society. This holds nutritionally and economically in the broadest sense, having an impact at the agricultural level as well as in materials and energy. As a global renewable resource, it is, therefore, crucial to consider starch in the context of the environmental challenges the world is facing. In the introduction of this book, these issues are addressed and considered through the principles of the bioeconomy and circular economy. Crucial market features complement these with respect to sources, producers, and trends. At the core of this bioeconomy are the biorefineries which sustainably transform biomass into food, feed, chemicals, materials, and bioenergy (fuels, heat, and power) generally through plant chemistry. Its sectors and industries have strong innovation potential

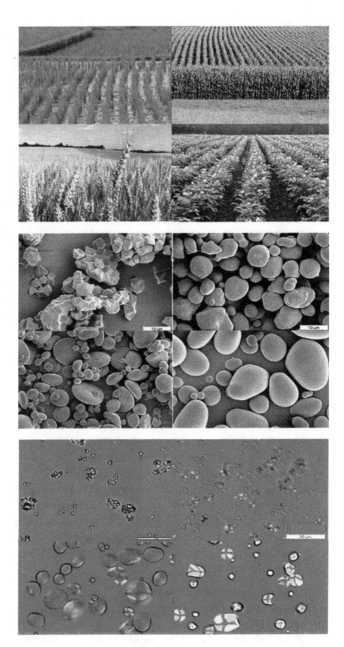

FIGURE 1   Rice, maize, wheat, and potato are starch-bearing crops, which are the main source of dietary energy for the world's population. Starches isolated from these sources (and others) show distinct characteristic granule morphology and dimensions. However, irrespective of their botanical sources, they all have a Maltese cross when observed in polarized light.

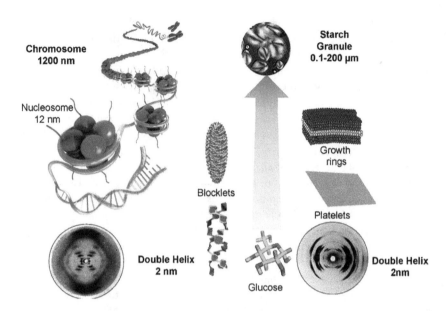

FIGURE 2   The universality of information transfer in living material starts with sugar–phosphate backbones arranged in an antiparallel double-helical structure having a repeat of 2 nm; this dictates the overall architectures of higher structures reaching dimensions of about 1200 nm. The universal way to store energy in plants through starch lies in the organization of the complex architecture of the polysaccharide with several well-organized strata that maintain the compacity over structures that develop on six orders of magnitudes. The architecture found in plants extends from the parallel-stranded double helices, having a 2 nm repeat, to the starch granule, up to 0.1 to 200 μm.

due to increasing demand for its goods and for their use of a wide range of scientific disciplines and industrial technologies.

Such innovation potential and technological achievements can only be developed based on sound scientific fundamentals and knowledge. Chapters 2, 3 and 4 address our present state of understanding of the starch granule, its biosynthesis, and biodegradation. Chapter 2 devoted to the structure of starch provides a key to the understanding of more than 50 years of research that have been dedicated to the elucidation of the structures of amylose and amylopectin. The reader will be introduced to the terminology that, over the years, has been designed to characterize the elements resulting from controlled starch hydrolysis. From these, various models have been proposed; however, none of them are entirely satisfactory in explaining starch hydrolysis, with different models still under scrutiny. Chapter 3 provides a detailed overview of starch biosynthesis. It covers the localization of synthesis in leaves, and storage tissues, and the enzyme reactions involved in starch synthesis. In addition to the core enzymes of the starch biosynthetic pathway, regulatory enzymes and non-catalytic proteins have been discovered that play a role in coordinating enzymes and enzyme complexes during starch synthesis. Although starch-biosynthetic enzymes are highly conserved, sufficient knowledge is still required for the rational modification of starch structure and properties in crops. Progress in engineering new starches with enhanced functionalities *in planta* has been empirical and slow. There is still much room to realize the full potential of starch. Chapter 4 devoted to starch degradation describes the biodegradation scheme occurring *in planta* with the transitory starch turnover in *Arabidopsis*. This is followed by a description of the current view on the pathway of starch degradation in leaves and cereal seeds. Chapter 5 is devoted to the description of the properties of starch and modified starches. It covers not only the physicochemical aspects but also starch digestibility. A comprehensive presentation of the different steps undergone during the ingestion, digestion, and absorption in humans ends with some considerations on resistant starches. Modifications of starch via physical, chemical, and biochemical processes yield a variety of substrates that constitute a formidable reservoir of products with functional properties that are considered indispensable ingredients in the industry. Chapter 6 gives an up-to-date picture of the worldwide production of starches with a particular emphasis on the main uses in the food and non-food industries, including the energy sector. Aside from the established manufacture of starch bioplastics, a new class of starch-based polymer nanocomposites offer a promising alternative to conventional plastic. Chapter 7 summarizes the elements and arguments in favor of a bio-based, circular economy focusing on (1) the sustainability of starch products; (2) thermoplastic starch-based renewable, biodegradable bioplastics; (3) starch for the production of bio-ethanol; (4) starch for new applications of ionic liquids; and (5) starch as a feedstock for advanced functional materials. Appreciation of the feasibility of such a vision lies in the sustainability of starch products under economic, environmental, and societal pillars.

The book "Starch in the Bioeconomy" is aimed at active or curious readers of starch science, technology, and economics. Built on a reliable and well-documented base of information, the book presents the paths that remain to be taken to decipher this still mysterious resource that has contributed so much to the rise of humanity. It also puts into perspective the enormous potential of starch and how it can contribute to developing a sustainable circular bioeconomy for the benefit of the economy, the environment, and society.

Serge Perez
Grenoble, April 2020

Authors

Jean-Luc Wertz holds degrees in chemical civil engineering and economic science from the Catholic University of Louvain, Louvain-la-Neuve, Belgium, as well as a PhD from the same university in applied science, specializing in polymer chemistry. He has had various international positions in R&D, including Spontex where he was the worldwide director of R&D. He holds several patents related to various products. In his last job before his retirement, he was a project manager in biomass valorization at ValBiom and worked more than eight years on bio-based products and biorefineries. He also wrote three books: *Cellulose Science and Technology* in 2010, *Lignocellulosic Biorefineries* in 2013, and *Hemicelluloses and Lignin in Biorefineries* in 2018.

Bénédicte Goffin holds a master's degree in Materials Engineering from the University of Louvain (Louvain-la-Neuve, Belgium) as well as a PhD in Chemistry from the University of Namur (Belgium). She joined Certech (Seneffe, Belgium) in 1998 as a Project Manager. With more than 20 years of experience in the field of polymer science, Bénédicte is leading several collaborative R&D projects at regional and European levels. This includes the scientific and financial management of public-funded projects such as cross-border projects (Interreg programme between France and Belgium). Through these various innovative projects, Bénédicte specialized more particularly on topics related to bio-based products, wood or natural fiber polymer composites, circular economy, and plastics recycling.

Acknowledgments

We express our gratitude to Serge Perez for his remarkable preface and his general input in this book, especially in the structure of starch.

Jean-Luc Wertz dedicates this book to his wife, Lydia, his two children, Vincent and Marie, and to his four grandchildren, Mathilda, Nicolas, Carolina, and Laura.

Bénédicte Goffin dedicates this book to his children, Mathieu and Guillaume.

Finally, we thank Allison Shatkin, Gabrielle Vernachio, and Joette Lynch of the Taylor & Francis Group for their efficient support and assistance during the publication of this work. We also thank Vijay Shanker of Codemantra for the production of this book.

1 Introduction

1.1 KEY FEATURES OF STARCH

When a plant absorbs carbon dioxide from the atmosphere and receives adequate sunlight and water, *chloroplasts* present in the plant's cells convert inorganic raw materials, water, and carbon dioxide into oxygen and glucose.[1] Glucose is stored in plant tissues for food and energy purposes. This process is termed as photosynthesis. Glucose is converted into starch, fats, and oils for storage. Starch is the most widespread and abundant storage carbohydrate in plants.[2]

Starch is a biopolymer that consists of two major components: amylose and amylopectin. Amylose, which forms up to 15%–35% of the granules in most plants, is primarily a linear polysaccharide with α-(1,4)-linked D-glucose units.[3] Amylopectin is a highly branched molecule with backbones made up of α-(1,4)-linked D-glucose units and branches made up of about 5% of α-(1,6)-linked D-glucose units.

Starch is synthesized in a *granular* form (Figure 1.1) in special organelles called plastids. Plastids store starch and are of two types: *chloroplasts* (Figure 1.2a), in which a temporary storage form is produced during photosynthesis, and *amyloplasts* (Figure 1.2b), in which long-term storage starch is produced.

Considering the great diversity in the morphology of starch granules, it is remarkable to observe that the internal architectural features of starch granules are shared universally among the plants and regardless of the plant organ.[5] The common features are growth rings, blocklets, and crystalline and amorphous lamellae. Most of the native starch granules exhibit a Maltese cross when observed under polarized light.

Starch granules are mainly found in seeds, roots, and tubers, as well as in stems, leaves, fruits, and even pollen.[3] Grain seeds, such as maize kernels,

FIGURE 1.1 Raw starch granules observed under a scanning electron microscope (SEM): (a) potato, (b) cassava, and (g) rice starches. The corresponding granules under polarized light are shown in the insets. The bottom figure shows the SEM image of *in situ* granules in a potato parenchyma cell. (Reproduced with permission from Serge Perez.)[4]

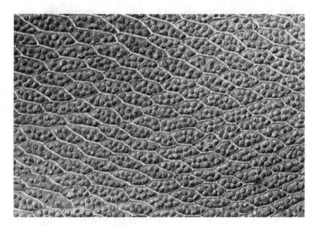

FIGURE 1.2A Chloroplasts (green spherical bodies) visible in the cells of *Bryum capillare*, a type of moss. (Reproduced with permission from Des Callaghan under the Creative Commons Attribution Share Alike 4.0 International license.)[6]

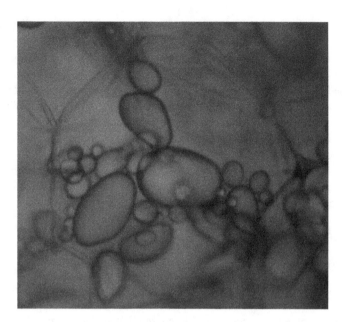

FIGURE 1.2B Amyloplasts in a potato cell. (Reproduced with permission from MNOLF under the Creative Commons Attribution Share Alike 3.0 Unported license.)[7]

contain up to 75% of starch. Biosynthesis of starch granules is initiated at the hilum, and the granule grows by *apposition* (i.e., by the gradual addition of layers of starch material).[8] The granules occur in all shapes and sizes (spheres, ellipsoids, polygons, platelets, and irregular tubules). They have diameters ranging from around 0.1 to 200 μm depending on their botanical origin. In some cereals, notably wheat and barley, there are two major populations of granules distinguished by their size, with the diameter cut-off being approximately 10 μm.[5] In barley, the average diameter of the large granules (also called A-granules) is between 15 and 19 μm depending on the variety, and the diameter of the small (B) granules is between 3.1 and 3.7 μm. Starch granules are densely packed with semi-crystalline structures, and the crystallinity varies from 15% to 45%, with a density of about 1.5 g/cm.[9]

We depend upon starch for our nutrition, exploit it in industry, and use it as a feedstock for the production of bioethanol. Starch is a major energy source for humans; it is degraded by digestive enzymes, called amylases, into a variety of disaccharides, trisaccharides, and oligosaccharides (dextrins). Amylases are a group of enzymes that hydrolyze glycosidic bonds present in starch and include α-amylase, β-amylase, and γ-amylase.[10] α-Amylase acts at random locations along the starch chain and breaks down long-chain carbohydrates to ultimately yield maltotriose and maltose (a disaccharide) from amylose, or maltose, glucose, and "limit dextrin" from amylopectin.[11] β-Amylase, which acts from the non-reducing end, catalyzes the hydrolysis of the second α-1,4-glycosidic bond and cleaves off two glucose units (maltose) at a time. γ-Amylase, also called glucoamylase, cleaves α-1,6-glycosidic linkages, as well as the last α-1,4-glycosidic linkages at the non-reducing end of amylose and amylopectin, to yield glucose.

A very important field in which enzymes have proved to be of great value over the past 15–20 years is the starch industry.[12] In the 1950s, fungal amylase was used in the production of specific types of syrup, i.e., those containing a range of sugars, which could not be produced by conventional acid hydrolysis. The real turning point was reached early in the 1960s when a glucoamylase enzyme was launched for the first time, which could completely break down starch into glucose. Within a few years, almost all glucose production was re-organized and enzyme hydrolysis was used instead of acid hydrolysis because of the benefits obtained, such as greater yield, higher degree of purity, and easier crystallization.

Here in this book, we describe starch as a key ingredient of the *bioeconomy* (bio-based economy). Starch is unique because it can be used in the production of food, materials (bio-based products), and energy.

1.2 BIOECONOMY

1.2.1 Bioeconomy and Circular Economy

Bioeconomy is defined as the production of renewable biological resources and the conversion of these resources and waste streams into value-added products such as food, feed, bio-based products, and bio-energy.[13] It includes agriculture, forestry, fisheries, and food, pulp, and paper production, as well as parts of chemical, biotechnological, and energy industries. Its sectors and industries have a strong innovation potential due to an increasing demand for its goods and their use of a wide range of sciences and industrial technologies.

Our current fossil-based economic system is under pressure from depleted supplies of raw materials, a changing climate, and population growth and aging.[14] Development of a bioeconomy can address several of these challenges. It is crucial for future generations.

The bioeconomy is circular by nature because carbon is sequestered from the atmosphere by photosynthesis in plants. After uses and reuses of products made from those plants, the carbon is cycled back as soil carbon or as atmospheric carbon once again (Figure 1.3).[15]

The circular economy focuses mainly on the efficient use of finite resources and ensures that those are used and recycled as long as possible.[16] The bioeconomy integrates the productions of renewable resources. It is the renewable part of the circular economy. The principle of the circular economy is thus complementary to the renewable character of the bioeconomy and must facilitate the recycling of carbon after efficient uses. The benefits from the circular economy will be truly felt if the bioeconomy is made to play its important role.

In the core of this bioeconomy are the biorefineries, which sustainably transform *biomass* into food, feed, chemicals, materials, and bioenergy (fuels, heat, and power) generally through plant chemistry. Biomass is defined as a material of biological origin, excluding the materials embedded in geologic formations and/or fossilized. In other words, it is any organic matter that is available on a renewable or recurring basis, including agricultural crops and trees, wood, wood wastes, wood residues, plants (including aquatic plants), grasses, plant residues, fibers, animal wastes, municipal wastes, and other waste materials.[17]

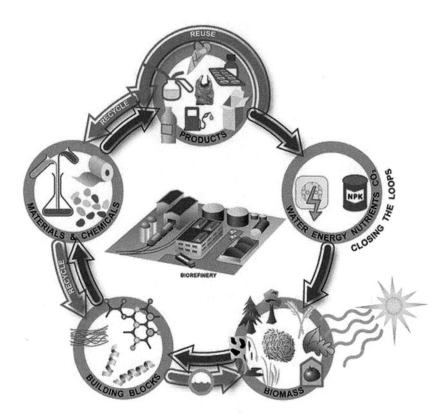

FIGURE 1.3 Bioeconomy as part of the circular economy. (Reproduced with permission from D. Carrez.)

1.2.2 Bioeconomy and Sustainability

Bioeconomy is the world's response to key environmental challenges the world is facing already today. It is meant to reduce the dependence on natural resources, transform manufacturing process, and promote sustainable production of renewable resources from land, fisheries, and aquaculture and their conversion into food, feed, fiber, bio-based products, and bioenergy, meanwhile growing new jobs and industries.[18]

The optimal production refers to modern concepts such as functional economy. A functional economy is one that optimizes the use (or function) of goods and services and thus the management of existing wealth (goods, knowledge, and nature).[19] The economic objective of the functional economy is to create the highest possible use value for the longest possible time while consuming as few material resources and energy as possible. This functional economy is therefore considerably more sustainable, or dematerialized, than the present economy. The model proposes to exchange the material goods by functional goods.

In 1989, the World Commission on Environment and Development has defined sustainable development as "development which meets the needs of current generations without compromising the ability of future generations to meet their own

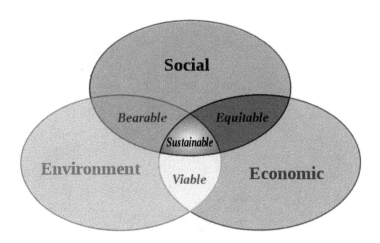

FIGURE 1.4   The three pillars of sustainability.[21,22]

needs".[20] Sustainability includes three pillars: environment, social, and economic. That definition explains without ambiguity that economic and social well-being cannot coexist with measures that destroy the environment (Figure 1.4).

In 2018, the bioeconomy sectors of the European Union (EU) already were worth Euro two trillion in annual turnover and accounted for more than 22 million jobs, i.e., approximately 9% of the total EU workforce.[23] It includes agriculture, forestry, fisheries, and food, pulp, and paper production, as well as parts of chemical, biotechnological, and energy industries.[24]

The bioeconomy offers a unique opportunity to address societal challenges such as food security, natural resource scarcity, fossil resource dependence, and climate change while achieving sustainable economic growth.[25]

In the bio-based economy, sometimes regarded as the non-food pillar of the bioeconomy, renewable biological resources instead of fossil resources are used as raw materials for the production of chemicals, materials, and fuels.[26] The bio-based economy involves the conversion of renewable feedstock (biomass and organic waste) into bio-based products.[27] Biotechnology plays a key role in the bio-based economy. Plant (green) biotechnology, on the one hand, is important for the primary production of biomass through the genetic improvement of crops. Industrial (white) biotechnology, on the other hand, is necessary for the conversion of biomass into various products, using microorganisms (fermentation) and their enzymes (*biocatalysis*).

1.2.3 Life Cycle Assessment

As environmental awareness increases, industries and businesses are assessing how their activities affect the environment.[28,29.] Companies have developed methods for assessing the environmental impacts associated with a product, process, or activity. One such tool is life cycle assessment (LCA).

LCA is a broadly accepted method that can be used to quantify the impacts along bioeconomy value chains. The LCA concept considers the entire life cycle of a product. It is a "cradle-to-grave" approach for assessing industrial systems. The "cradle-to-grave" approach starts with the gathering of raw materials from the earth to create a product and ends at the point when all materials are returned to the earth. By contrast, "cradle-to-cradle" is a model of industrial systems in which a material flows cyclically in appropriate, continuous biological or technical nutrient cycles. All waste materials are efficiently re-incorporated into new phases of production and usage, i.e., "waste equals food".[30]

The term "life cycle" refers to the major activities in the course of a product's lifespan from its manufacture, use, and maintenance to its final disposal, including the raw material acquisition for its manufacture.[29]

LCA is part of the ISO 14000 series of standards on environmental management.[31–33] The series provides principles, framework, and methodological requirements for conducting LCA studies.[34] The LCA framework consists of four components: goal and scope definition, inventory analysis, impact assessment, and interpretation.

LCA studies on starch plastics have been performed. Starch plastics are developed for their bio-based origin and potential biodegradability.[35] They are shown to enable reductions in greenhouse gas (GHG) emissions and non-renewable energy use, but they have higher eutrophication potential and require more agricultural land use compared to common petrochemical plastics (on a same weight basis).

1.3   STARCH MARKET—SOURCES, PRODUCERS, AND TRENDS

Starch is the most important, abundant, digestible food polysaccharide and is therefore a major source of energy in human diets.[36] Starches are the major storage polysaccharides in foods of plant origin.[37] Common food starches are derived from cereals (such as wheat, corn, rice, and barley) and tubers and roots (such as potato and cassava).

The global starch market is poised to register a compound annual growth rate (CAGR) of 5.85% for the period from 2018 to 2023.[38] The market is competitive and driven by an increase in the trend of "health and wellness" and growing consumer demand for all natural ingredients. This trend includes weight-management foods and functionally diverse ingredients. Increasing advancements in microencapsulation have widened the options for the starch industry.

Worldwide dry starch production was estimated to be more than 64 million tons in 2009 and almost 75 million tons were expected by 2012.[38] The annual production of primary starch sources in 2009 was estimated to be 46.1 million tons of corn, 9.1 million tons of cassava, 5.15 million tons of wheat, and 2.45 million tons of potato.

Cereal starch comes mainly from corn and wheat.[38] The United States is the largest producer of corn starch in the world. In Western Europe, 46% of produced starch is from corn, 36% from wheat, and 19% from potato, whereas in North America, the production of starch is based almost completely on corn. The Asian corn starch sector, particularly in China, is growing consistently. Because of their lower moisture content, cereals have longer storage times and extraction of starch from cereals is easier and faster than from roots and tubers. Europe is responsible for 80% of potato starch production.

Major industrial companies in the starch business are Archer Daniels Midland, Agrana, Avebe, Beneo, Cargill, Grain Processing Corp., Ingredion, National Starch Food Innovation, Roquette Frères, and Tate & Lyle-Tereos Syral.[39]

The European starch industry produces over 600 products, from native starches to physically or chemically modified starches to liquid and solid sweeteners.[39] The versatility of starch products is such that they are used as ingredients and functional supplements in a vast array of food, non-food, and feed applications. From 74 starch production facilities in 20 of the 28 EU member states, the European Starch Industry today produces 10.7 million tons of starch each year from wheat, maize, and potatoes. The consumption of starch and starch derivatives in the EU was 9.3 million tons in 2016. Of the 9.3 million tons of starch and starch derivatives consumed in the EU, from a product point of view, 26% are native starches, 19% are modified starches, and 55% are starch sweeteners, and from an application point of view, 61% are in food, 1% is in feed, and 38% are in non-food applications, mainly paper making.

1.4 STARCH IN PLANT CELLS

1.4.1 TRANSITORY AND STORAGE STARCH

Starch is synthesized in the *plastids*, which are of two types: (1) chloroplasts in leaves and (2) amyloplasts in the starch-storing tissues of staple crops. Based on its biological functions, starch is often categorized into two types: transitory starch found in chloroplasts (green plastids) and storage starch found in amyloplasts (non-photosynthetic plastids).[40] Chloroplasts are the site of photosynthesis and, among others, of starch synthesis.[41] Plastids in non-photosynthetic tissues are the site of syntheses of fatty acid, starch, and amino acid.[42]

- The starch that is synthesized in the leaves directly from the *photosynthates* during the day is typically defined as transitory starch, because it is degraded in the following night to sustain metabolism, energy production, and biosynthesis in the absence of photosynthesis.
- The starch in non-photosynthetic tissues, such as seeds, stems, roots, and tubers, is generally stored for longer periods and regarded as storage starch. Remobilization of starch takes place during germination, sprouting, or re-growth, when photosynthesis cannot meet the demand for energy and biosynthesis. It is this storage starch that we consume as our food and extract for industrial uses—it can account for 70% to 80% of the dry weight in wheat grains and cassava roots.[43]

Starches from different botanical sources vary in terms of their functional properties and thus in their end-uses. This variation stems from differences in the structure of starch, such as the size of starch granules, their composition, and molecular architecture of the constituent polymers.

1.4.2 HOW CELLS OBTAIN ENERGY FROM STARCH

To be utilized for the production of energy, starch, like other food molecules, must first be converted into a biochemically accessible form. The energy is derived from the chemical bond energy in starch molecules.[44] Starch is oxidized in small steps to carbon dioxide and water. The complete aerobic oxidation of glucose is coupled to the synthesis of 36 molecules of ATP (adenosine triphosphate) from ADP (adenosine diphosphate) (Equation 1.1):

$$C_6H_{12}O_6 + 6O_2 + 36P_i^{2-} + 36ADP^{3-} + 36H^+ \rightarrow 6CO_2 + 36ATP^{4-} + 42H_2O \quad (1.1)$$

Starch must be broken down into glucose before the cells can use it, either as a source of energy or as building blocks for other molecules. This preliminary stage in the breakdown of starch is called *digestion*. The large polymeric molecules are broken down during digestion into glucose through the action of enzymes. After digestion, glucose enters the *cytosol* of the cell, where its gradual oxidation begins: *glycolysis*, the initial stage of glucose metabolism, does not involve molecular oxygen and produces a small amount of ATP and the three-carbon compound pyruvate. In aerobic cells, pyruvate formed in glycolysis is transported into the mitochondria, where it is oxidized by oxygen to carbon dioxide. Through *chemiosmotic* coupling, the oxidation of pyruvate in the mitochondria generates the bulk of the ATP produced during the conversion of glucose to carbon dioxide. The oxidation of pyruvate involves the citric acid cycle, in which *acetyl CoA* derived from pyruvate is modified to produce energy precursors in preparation for the next step, and the oxidative phosphorylation, in which ADP is transformed into ATP.

1.5 STARCH BIOREFINERIES

1.5.1 BIOREFINERY CONCEPT

Simultaneously resolving energy security and environmental concerns is a key challenge for policy makers today.[45] The world is facing a fast-growing human population and the consequent growing demand for food, water, and energy.[46,47] Furthermore, anthropogenic climate change is a severe threat to mankind and requires a significant reduction of our current GHG emissions to avoid detrimental consequences for the planet. Only the use of new technologies will allow us to bridge the gap between economic growth and environmental sustainability progressively. These technologies will transform the world economy from the one based on fossil resources to the one based on renewable resources such as biomass while improving the sustainable production of energy, fuels, chemicals, and materials.[48–50] The major drivers for a bioeconomy are independence from fossil energy sources, climate change abatement, and economic growth through new value chains.[51]

Biorefineries that convert biomass into chemicals and fuels were identified as a potential solution to mitigate the threat of climate change and to meet the growing demand for energy and non-energy products.[49] Generally, one can make a distinction between first- and second-generation biorefineries.[52]

First-generation biorefineries, which are already well implemented around the world, convert edible biomass, such as corn, wheat, and sugar cane/beet, into energy and non-energy products. As production levels have increased, along with human populations, concerns about competition with food needs have arisen. Land use changes, whether direct or indirect, are one of the most important consequences of bioenergy production.[53] Most biofuels today use feedstock grown on land that is suitable for food, feed, and material production. An increase in biofuel production could therefore lead to cropland expansion in one of the two ways: (1) directly, when new cropland is created for the production of biofuel feedstock, and (2) indirectly, when existing cropland is used for biofuel feedstock production, forcing food, feed, and materials to be produced on new cropland elsewhere. Nevertheless, over the past 30 years, these first-generation feedstocks have paved the way for more sustainable production of biofuels.[54]

Second-generation biorefineries use lignocellulosic feedstocks, such as wood, corn stover, and bagasse, and convert them into energy and non-energy products. Their objective is to optimize the valorization of all plant components. Most lignocellulosic biorefineries are expected to be ready for large-scale commercial production in a few years.

1.5.2 FEEDSTOCK AND GENERATED PRODUCTS

Common starch-containing feedstocks include corn, wheat, potato, pea, barley, and cassava.

Corn, wheat, potato, and pea are the raw materials used by Roquette, a leader in starch production in Europe, in their Lestrem Starch Biorefinery (Figure 1.5).[55] Currently, the Lestrem Starch Biorefinery converts 7,000 tons of grains each day and markets more than 700 products from starch for food, feed, and industrial applications (paper, pharmacy, chemicals, cosmetics, and so on).

Tereos Syral is the third European starch producer and transforms three million tons of wheat and corn into starches and derivatives, vegetable proteins, alcohol, and bioethanol in its six plants.[56] These products are used for human and animal nutrition, pharmaceuticals and personal care, and in the production of paper and corrugated board, chemicals, plastics, and biofuels.

FIGURE 1.5   Raw materials used by Roquette. (Reproduced with permission from Roquette.)

1.5.3 STARCH BIOREFINERY SCHEME

1.5.3.1 Gelatinization

Starch is an important feedstock for industrial applications. Starch granules are quite resistant to penetration by both water and hydrolytic enzymes (e.g., α-amylase) due to the formation of hydrogen bonds within the same molecule and with other neighboring molecules. Heating an aqueous starch suspension weakens these inter- and intra-hydrogen bonds, causing swelling of the starch granules due to absorption of water. This process is commonly called *gelatinization* because the solution formed has a gelatinous, highly viscous consistency.

1.5.3.2 Liquefaction and Saccharification

Gelatinized starch is converted into glucose and maltose in the industrial process by two enzymes: α-amylase and glucoamylase. First, the starch polymer is hydrolyzed by α-amylase to shorter chains called *dextrins* in a process known as *liquefaction* because

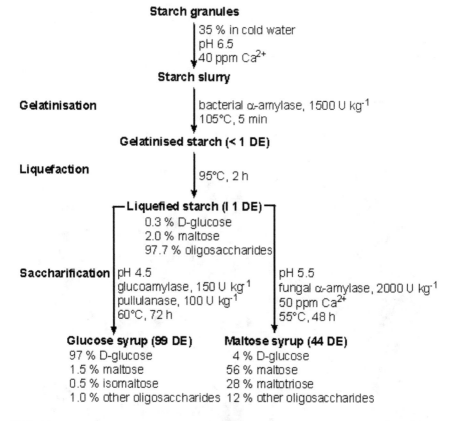

FIGURE 1.6 Using enzymes in the processing of starch; typical conditions are given.[59] DE, dextrose equivalent, represents the percentage hydrolysis of the glycosidic linkages present (glucose has a DE of 100, maltose has a DE of ~50, and starch has a DE of effectively 0). In general, the DE of liquefied starch suitable for saccharification is ~10 for glucose production and 0.5–5 for maltose production. (Reproduced with permission from Martin Chaplin.)[60]

the breakdown of polymers yields a thinner solution. Finally, the dextrins are degraded to glucose and maltose by glucoamylase and α-amylase in a process called *saccharification*, defined as the release of simple sugars from a polymer (Figure 1.6).[57,58]

A novel raw starch-digesting glucoamylase from the fungus *Penicillium oxalicum* has been shown to allow an efficient hydrolysis of raw starch and ethanol fermentation.[61] Raw starch-digesting glucoamylases are capable of directly hydrolyzing raw starch to glucose at low temperatures, which significantly simplifies processing and reduces the cost of producing starch-based products.

1.5.3.3 Fermentation

Saccharification is followed by *fermentation*. During fermentation, the *yeast* transforms sugars into ethanol and carbon dioxide. The yeast used in the starch biorefinery is *Saccharomyces cerevisiae*, which is commonly used in the production of beverage alcohol (Figure 1.7).[62]

In ethanol fermentation, (1) one glucose molecule breaks down into two pyruvates (Figure 1.8). The energy from this exothermic reaction is used to bind the inorganic phosphates to ADP and convert NAD+ into NADH. (2) The two pyruvates are then broken down into two acetaldehydes and give off two carbon dioxide molecules as a by-product. (3) The two acetaldehydes are then converted into two ethanol molecules by using the H+ ions from NADH, converting NADH back into NAD+.

An important process development made for starch enzymatic hydrolysis is the introduction of simultaneous saccharification and fermentation (SSF) process.[65] Various SSF processes were developed and evaluated using the bacterium

5 μm

FIGURE 1.7 *Saccharomyces cerevisiae*. (Reproduced from Mogana Das Murtey and Patchamuthu Ramasamy under the Creative Commons Attribution Share Alike 3.0 Unported license via Wikimedia Commons.)[63]

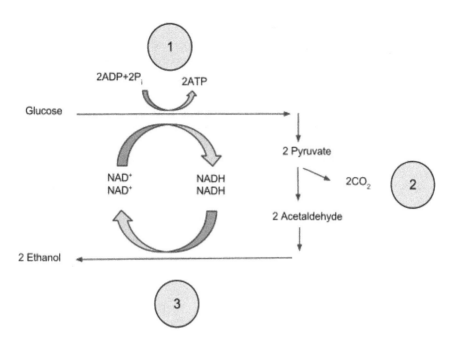

FIGURE 1.8   Ethanol fermentation. (Reproduced from David Carmack under the Creative Commons Attribution Share Alike 3.0 Unported license via Wikimedia Commons.)[64]

Zymomonas mobilis.[66] Compared with a two-step process involving separate stages, the SSF reduced the total process time by half.

1.5.3.4   Status and Potential

Starch biorefineries efficiently valorize the raw materials, energy, and water during the production process, which contributes to the competitiveness of the EU starch industry.[67] The EU starch industry processes every part of the plant and produces minimal waste; less than 1% is not valorized. It is worth knowing that in 2015, the EU starch industry processed about 23 million tons of agricultural raw materials, roughly split equally among wheat, corn, and potatoes, transformed into 11 million tons of starch.

One promising development is in the whole crop biorefinery, where starch grain and straw fractions are processed into a portfolio of end products.[68] It encompasses dry or wet milling and consequent fermentation and distilling of grains.

1.6   BIO-BASED PLASTICS INCLUDING THERMOPLASTIC STARCH

Increased use of biomass resources for manufacturing plastics is effective in reducing global warming and the depletion of fossil resources.[69] According to ISO, a plastic is "a material which contains as an essential ingredient a high polymer and which, at some stage in its processing into finished products, can be shaped by flow".[70] Bio-based plastics (or simply bioplastics) refer to plastics that contain materials **wholly or partly** of biogenic origin.

Bio-based plastics include thermoplastic starch, which is a modified starch obtained by using plasticizers. Some bio-based polymers, with the composition of bio-based plastics, have a structure identical to fossil-based polymers, such as polyethylene and polyethylene terephthalate, while others have a new structure, such as polylactic acid (PLA).

Biopolymers can be bio-based and biodegradable, or bio-based and non-biodegradable, or non-bio-based and biodegradable, or non-bio-based and non-biodegradable (Table 1.1).[71] Plastics are a shortened version of thermoplastics, different from thermosets. Biopolymers include bio-based plastics such as PLA but also include natural polymers such as starch and cellulose. PLA is a biopolymer and a bio-based plastic, while starch and cellulose are biopolymers but not bio-based plastics.

Figure 1.9 from Nova Institute shows the production capacities of bio-based polymers from 2019 to 2024.[72]

TABLE 1.1
Categories of Biopolymers[72]

Origin	Biodegradability	Example	Meaning of bio
Bio-based	Biodegradable	Starch, polylactic acid, polyhydroxyalkanoate	Bio-based and biodegradable
Bio-based	Non-biodegradable	Polyethylene from sugar cane	Bio-based
Fossil	Biodegradable	Polycaprolactone and aliphatic polyesters in general	Biodegradable
Fossil	Non-biodegradable	Polyether ether ketone	Biocompatible

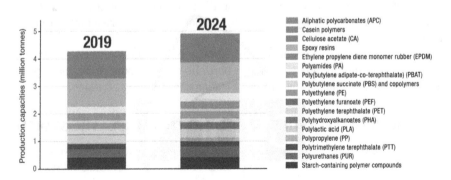

FIGURE 1.9 Global production capacities of bio-based polymers in 2019 and 2024. (Reproduced with permission from the Nova Institute.)[72]

The installed capacity was 4.3 million tons in 2019 and should increase to 4.9 million tons in 2024, which indicates an expected CAGR of about 3%. In 2019, the global production volume of bio-based polymers was 3.8 million tons, which is 1% of the production volume of fossil-based polymers.

1.7   STRUCTURE OF THE BOOK

Chapter 1 is an introduction to the book and includes key features of starch, bio-economy, starch market, starch in plant cells, starch biorefineries, and thermoplastic starch as part of bio-based plastics.

Chapter 2 is devoted to the structure of starch and explains starch components and the starch granule.

Chapter 3 provides a detailed overview of starch biosynthesis and includes localization of synthesis and enzyme reactions involved.

Chapter 4 describes the enzymatic degradation of starch and includes transitory turnover in *Arabidopsis*, pathway of degradation in leaves, and pathway in cereal seeds.

Chapter 5 presents the properties of starch and modified starches, including physicochemical properties, starch digestibility, unconventional starches, physical and chemical modifications of starch, and thermoplastic starch.

Chapter 6 is about applications of starches in the bioeconomy and includes starch industry, influence of starch structure on its function, applications in the food industry, and applications in the non-food industry.

Chapter 7 gives perspectives of starch in the bioeconomy and includes sustainability of starch products, thermoplastic starch, starch ethanol, and advanced starch applications.

REFERENCES

1. S. Petersen, *Sciencing, What Are the Functions of Starch in Plant Cells?* 2018 in https://sciencing.com/functions-starch-plant-cells-5089163.html
2. S.C. Zeeman, J. Kossmann and A.M. Smith, *Annu. Rev. Plant Biol.*, 61, 209, 2010 in https://www.ncbi.nlm.nih.gov/pubmed/20192737
3. A. Sarkar and S. Perez, A database of polysaccharide 3D structures, *Starch*, 2012 in http://polysac3db.cermav.cnrs.fr/discover_starch.html
4. D.J. Gallant, B. Bouchet, A. Buleon, S. Perez, *Eur. J. Clin. Nutr.*, 46:S3, 1992 in https://pubmed.ncbi.nlm.nih.gov/1330527/
5. E. Bertoft, *Agronomy*, 7, 56, 2017 in http://www.mdpi.com/2073-4395/7/3/56/htm
6. *Chloroplasts* in https://en.wikipedia.org/wiki/Chloroplast
7. *Amyloplasts* in https://en.wikipedia.org/wiki/Amyloplast
8. M. Sjoo, L. Nilsson (Eds), Starch in Food, 2nd edition, *Structure, Function and Applications*, Woodhead Publishing, 2017 in https://www.elsevier.com/books/starch-in-food/sjoo/978-0-08-100868-3
9. S. Perez and E. Bertoft, *Starch/Stärke*, 62, 389, 2010 in https://onlinelibrary.wiley.com/doi/abs/10.1002/star.201000013 and in http://agris.fao.org/agris-search/search.do?recordID=US201301876336
10. H. Taniguchi, Y. Honnda, *Encyclopedia of Microbiology*, 3rd edition, 2009 in https://www.sciencedirect.com/topics/neuroscience/amylase

11. *Amylase* in https://en.wikipedia.org/wiki/Amylase
12. Maps Enzymes Limited, 2010 in http://www.mapsenzymes.com/history_of_enzymes.asp
13. Council of the European Union, 5757/18, AGRI 52 in http://www.consilium.europa.eu/media/32637/revision-of-the-eubioeconomy-strategy-and-the-role-of-the-agricultural-sector.pdf
14. Research Institute for Agriculture, Fisheries and Food, *Flanders, State of the Art* in https://www.ilvo.vlaanderen.be/language/en-US/EN/Press-and-Media/All-media/articleType/ArticleView/articleId/4528/How-can-we-accelerate-the-transition-from-a-fossil-economy-to-a-bio-economy-A-new-organization-system-for-targeted-innovation.aspx
15. Bio-based Industries Consortium, *Bioeconomy: Circular by Nature*, 2015 in https://biconsortium.eu/news/bioeconomy-cicular-nature
16. D. Carrez and P. Van Leeuwen, *Bioeconomy: Circular by Nature*, 2015 in http://biconsortium.eu/sites/biconsortium.eu/files/downloads/European_Files_september2015_38.pdf
17. Expert Group for the Review of the Bioeconomy Strategy and Its Action Plan, *Annex 4, Final Report, Glossary of Terms and Definitions*, 2017 in http://ec.europa.eu/transparency/regexpert/index.cfm?do=groupDetail.groupDetailDoc&id=36187&no=5
18. European Commission in http://ec.europa.eu/programmes/horizon2020/
19. W.R. Stahel *The Functional Economy: Cultural and Organizational Change* in http://www.product-life.org/en/archive/the-functional-economy-cultural-and-organizational-change
20. United Nations Economic Commission for Europe, UNECE in 2004–2005, *Sustainable Development-Concept and Action* in http://www.unece.org/oes/nutshell/2004-2005/focus_sustainable_development.html
21. Green Planet Ethics in http://greenplanetethics.com/wordpress/sustainable-development-can-we-balance-sustainable-development-with-growth-to-help-protect-us-from-ourselves
22. *Sustainable Development* in http://en.wikipedia.org/wiki/File:Sustainable_development.svg
23. European Commission, *Horizon 2020*, 2018 in https://www.ecologic.eu/sites/files/event/2018/bio-based-products_flyer_a5_online.pdf
24. European Commission in http://europa.eu/rapid/press-release_IP-12-124_en.htm
25. European Commission in http://ec.europa.eu/research/bioeconomy/press/newsletter/2012/02/sustainable_economy/index_en.htm
26. BEsustainable in http://www.besustainablemagazine.com/cms2/sustainable-use-of-and-value-creation-from-renewable-resources-in-a-biobased-economy-in-flanders/
27. Essenscia in http://www.essenscia.be/Upload/Docs/Bio.be-strategybiobasedeconomy.pdf
28. Scientific Applications International Corporation (SAIC), *Life Cycle Assessment: Principles and Practice*, EPA/600/R-06/060, 2006 in http://www.epa.gov/nrmrl/std/lca/pdfs/600r06060.pdf
29. Australian Life Cycle Assessment Society in http://www.alcas.asn.au/intro-to-lca
30. M. Braungart, *C2C Design Concept* in http://braungart.epea-hamburg.org/en/content/c2c-design-concept
31. United States Environment Protection Agency in http://www.epa.gov/nrmrl/std/lca/lca.html
32. ISO, *ISO 14000 Essentials*, 2011 in http://www.iso.org/iso/iso_14000_essentials
33. ISO, *ISO 14040: 2006* in http://www.iso.org/iso/catalogue_detail?csnumber=37456
34. Dantes website, *More about LCA* in http://www.dantes.info/Tools&Methods/Environmentalassessment/enviro_asse_lca_detail.html
35. M. L. M. Broeren, L. Kuling, E. Worrell and L. Shen, *ScienceDirect*, 127, 246, 2017 in https://www.sciencedirect.com/science/article/pii/S0921344917302793

36. Food Science and Technology, *Starch Content in Food*, 2014 in https://foodscience-techn.blogspot.com/2014/07/starch-content-in-food.html

37. A.C. Bertolini (Ed.), *Starches: Characterization, Properties and Applications*, CRC Press, 2009 in https://www.amazon.fr/Starches-Characterization-Applications-Andrea-Bertolini/dp/1420080237

38 40 Business Wire, 2018 in https://www.businesswire.com/news/home/20180504005332/en/Global-Food-Starch-Market-Analysis-Growth-Trend

39. Starch.Eu in https://www.starch.eu/the-european-starch-industry/

40. B. Pfister and S.C. Zeeman, *Cell Mol. Life Sci.*, 73, 2781, 2016 in https://www.ncbi.nlm.nih.gov/pmc/articles/PMC4919380/

41. P.A. Salome, M. Oliva, D. Weigel and U. Kramer, *EMBO J.*, 32, 511, 2012 in https://www.ncbi.nlm.nih.gov/pmc/articles/PMC3579136/

42. M.J. Emes and H.E. Neuhaus, *J. Exp. Bot.*, 48, 1995, 1997 in https://academic.oup.com/jxb/article/48/12/1995/681639

43. J.R. Lloyd and J. Kossmann, *Curr. Opin. Plant Biol.*, 32, 143, 2015 in https://www.sciencedirect.com/science/article/pii/S0958166914002146

44. B. Alberts et al., *Molecular Biology of the Cell*, 4th edition, Garland Science, 2002 in https://www.ncbi.nlm.nih.gov/books/NBK26882/

45. International Energy Agency, *Energy Security and Climate Policy*, 2007 in http://www.iea.org/publications/free_new_Desc.asp?PUBS_ID=1883

46. World Economic Forum, *The Future of Industrial Biorefineries*, 2010 in http://www3.weforum.org/docs/WEF_FutureIndustrialBiorefineries_Report_2010.pdf

47. OECD, *Green Growth Studies, Energy*, 2011 in http://www.oecd.org/dataoecd/37/42/49157219.pdf

48. Wageningen UR, *Biobased Economy across the Board*, 2011 in http://www.themabiobasedeconomy.wur.nl/UK/newsagenda/news/Biobased_Economy_across_the_board.htm

49. M. Bonaccorso, *Inside the World Bioeconomy,* 2014, Il Bioeconomista, in http://www.lulu.com/shop/mario-bonaccorso/inside-the-world-bioeconomy/paperback/product-21878056.html

50. United Nations, *RIO+20, Corporate Sustainability Forum*, 2012, in http://csf.compact4rio.org/events/rio-20-corporate-sustainability-forum/custom-114-251b87a2deaa4e-56a3e00ca1d66e5bfd.aspx

51. European Biogas Association, *EBA's Position on Bioeconomy*, http://european-biogas.eu/wp-content/uploads/2013/08/2014-08-EBA-position_bio-economy.pdf

52. International Energy Agency, *Sustainable Production of Second-Generation Biofuels*, 2010 in http://www.iea.org/papers/2010/second_generation_biofuels.pdf

53. Ecofys, *Iiasa and E4tech*, 2015 in https://ec.europa.eu/energy/sites/ener/files/documents/Final%20Report_GLOBIOM_publication.pdf

54. S.R. Hughes, W.R. Gibbons, B.R. Moser and J.O. Rich, *Intech* in http://cdn.intechopen.com/pdfs/42206/InTech-Sustainable_multipurpose_biorefineries_for_third_generation_biofuels_and_value_added_co_products.pdf

55. Lestrem Starch Biorefinery in https://biorrefineria.blogspot.com/2015/05/lestrem-starch-biorefinery.html

56. Tereos Syral in http://www.bio-based.eu/ibib/pdf/64.pdf

57. M.E. Carr, L.T. Black and M.O. *Biotechnol. Bioeng.*, 24, 2441, 1982 in https://www.ncbi.nlm.nih.gov/pubmed/18546215

58. ICM Inc, *Novozymes North America* in https://patents.google.com/patent/US20080009048

59. M. Chaplin, *Enzyme Technology, The Use of Enzymes in Starch Hydrolysis*, Cambridge University Press, 2014 in http://www1.lsbu.ac.uk/water/enztech/starch.html

60. The Amylase Research Society of Japan (Ed.),*Handbook of Amylases and Related Enzymes*,PergamonPress,1988inhttps://www.sciencedirect.com/book/9780080361413/handbook-of-amylases-and-related-enzymes

61. Q.S. Xu, Y.S. Yan and J.X. Feng, *Biotechnol. Biofuels*, 9, 216, 2016 in https://www.ncbi.nlm.nih.gov/pmc/articles/PMC5069817/

62. S. Minteer (Ed), *Alcoholic Fuels*, CRC Press, 2006 in https://www.crcpress.com/Alcoholic-Fuels/Minteer/p/book/9780849339448

63. *Saccharomyces cerevisiae*, https://en.wikipedia.org/wiki/Saccharomyces_cerevisiae

64. *Ethanol Fermentation* in https://en.wikipedia.org/wiki/Ethanol_fermentation

65. R.C. Ray and S.K. Naskar, *Dynamic Biochemistry, Process Biotechnology and Molecular Biology*, Global Science Books, 2008 in http://www.globalsciencebooks.info/Journals/DBPBMB.html

66. C.H. Kim and S.K. Rhee, *Process Biochem.*, 28, 331, 1993 in https://www.sciencedirect.com/science/article/pii/0032959293850073

67. Starch.Eu in https://www.starch.eu/blog/2017/05/04/starch-europe-position-cap-post-2020/

68. E. De Jong and G. Jungmeier, *Industrial Biorefineries and White Biotechnology* in https://www.iea-bioenergy.task42-biorefineries.com/upload_mm/3/7/5/cf7aa6b6-2140-46f2-b4ca-455f5c3eb547_de%20Jong%202015%20Biorefinery%20Concepts%20in%20Comparison%20to%20Petrochemical%20Refineries%20Book%20Chapter.pdf

69. ISO/TC61 in https://committee.iso.org/sites/tc61/home/projects/ongoing/ongoing-1.html

70. ISO 472: 2013 in https://www.iso.org/obp/ui/#iso:std:iso:472:ed-4:v1:en

71. European Commission, *Taking Bio-based from Promise to Market*, 2009 in https://www.iwbio.de/fileadmin/Publikationen/IWBioPPublikationen/bio_based_from_promise_to_market_en.pdf

72. nova-Institute, 2020 in http://bio-based.eu/downloads/bio-based-building-blocks-and-polymers-global-capacities-production-and-trends-2019-2024/

2 Structure of Starch

2.1 STARCH COMPONENTS

Starch, the carbohydrate reserve of plants, consists of two glucose polymers: amylopectin and amylose, which together form insoluble, semi-crystalline starch granules.[1-3] The molecular structure of amylose is comparatively simple as it consists of D-glucose units connected through α-1,4-*glycosidic linkages* to long chains with a few α-1,6 branches (Figure 2.1).[4] The character α or β of the anomeric configuration of the glycosidic linkages results from the anomeric form of the monosaccharide unit linked at C1. The α-character, like in starch, involves the formation of axial linkages perpendicular to the mean plane of the ring. The β-character, like in cellulose, involves the formation of equatorial linkages in the mean plane of the ring and alternating units rotated through 180°. Glucose shows a $^4C_1$ *chair conformation.*[5] Amylopectin, which is the major component, has the same basic molecular structure as amylose, but it has considerably shorter chains and a lot of α-1,6 branches.[1] Short, external chains of amylopectin are clustered and form double helices, which crystallize and contribute to the semi-crystalline nature of the starch granules.[1]

Clustered amylopectin side chains and amylose chains are organized in the helix conformation that subsequently forms crystalline structures, which can be divided into three types: A, B, and C.[6-8] The three structures correspond to the naturally occurring A-, B-, and C-starches. In A-type crystalline starch, glucose helices are packed densely, whereas in B-type crystalline starch, those are packed less densely, leaving room for water molecules in between the branches. C-type crystalline starch consists of a combination of A- and B-type crystallinity.

2.1.1 AMYLOSE

2.1.1.1 Molecular Structure

Amylose is the minor, linear, or slightly branched component of starch. It frequently forms a helix and is thought to intertwine even through the several layers of amylopectin.[9] Generally, branched amylose molecules are larger than their linear counterparts; however, the average chain lengths are shorter than the single chain in linear amylose. The molecular size of amylose also varies between starches (Table 2.1). Potato amylose is among the largest reported, whereas cereal amyloses are smaller. The proportion of branched amylose is estimated by the degradation of the amylose molecules with the enzyme β-amylase. The proportion of branched amylose molecules varies among plants: the molar fractions of branched amylose in wheat, rice, and maize are 0.27, 0.31, and 0.44, respectively, whereas in sweet potato, it is 0.70 (Table 2.1).

Table 2.2 gives the amylose content in eight additional starches not included in Table 2.1.

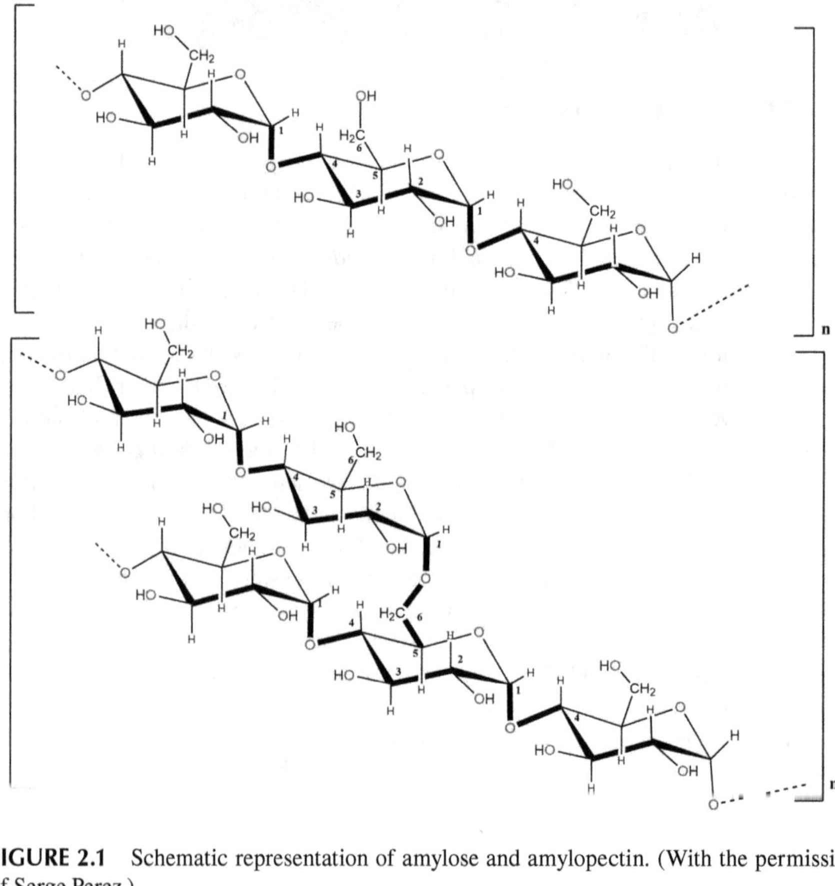

FIGURE 2.1 Schematic representation of amylose and amylopectin. (With the permission of Serge Perez.)

TABLE 2.1
Amylose Content and Structure in Selected Starches[1,10]

Source	Amylose (%)	DP_n[a]	$N_{branched}$ (%)[b]	$NC_{branched}$[c]
Wheat	17–34	980–1570	26–44	12.9–20.7
Barley	22–27	1,220–1,680	21–45	6.1–13.8
Rice	17–29	920–1,110	31–69	5.7–9.7
Maize	20–28	690–960	44–48	5.3–5.4
Sweet potato	19–20	3,280	70	13.6
Potato	25–31	4,920–6,340	n.a.	n.a.

[a] Number average degree of polymerization.
[b] Molar fraction of branched amylose molecules.
[c] Average number of chains in branched molecules.

TABLE 2.2
Amylose Content in Starches Not Mentioned in Table 2.1[a]

Source	Amylose (%)
Rye	26–30
Oat	18–29
Round pea	30–40
Sorghum	22–30
Wrinkled pea	60–76
Cassava	16.8–21.5
Waxy maize[a]	<2
Amylomaize	57–75

[a] Waxy maize contains no or very little amylose.

Branched amyloses also possess short chains with lengths normally attributed to amylopectin (i.e., with $DP < 100$).[1] The relative amount of these chains by weight is very low and often below the detection limit. On a number (molar) basis, however, these chains even predominate in the amylose. In contrast, the organization of the short chains in amylose is different from that in amylopectin, because their size-distribution is different from that generally found in amylopectin.

Linear amylose, like cellulose, has one reducing end containing an unsubstituted hemiacetal and one non-reducing end containing an additional hydroxyl group at C4.[11,12]

2.1.1.2 Amylose in the Granule

Three main hypotheses for the location and state of amylose have been put forward.[13] The first hypothesis suggested that amylose is laid down tangentially to the radial orientation of the amylopectin to minimize the amylose/amylopectin helical interactions. The other two hypotheses advocate radial deposition of the amylose, either in bundles or as individual chains, randomly interspersed among the amylopectin clusters in both the crystalline and semi-crystalline regions. Of the three hypotheses, the third hypothesis appears to be the most sustainable, because it has been demonstrated that amylose chains do not crosslink to one another (thereby ruling out the second hypothesis) but do crosslink to amylopectin chains.[14] The currently accepted model of amylose location in starch granules is thus based on individual, radially oriented amylose chains that are randomly distributed among the radial amylopectin chains.[15]

The preferential location of amylose within the peripheral or internal parts of the granules is still an open question even if the polymer is thought to be in the amorphous regions of the starch granules.[1] Some authors, including Perez, stated that there is an enrichment of amylose toward the periphery of the granule and that the amylose found near the surface of the granule has a smaller chain length than amylose located nearer the center of the granule.[11] In particular, Jane et al.[16,17] revealed that in both potato and maize starch granules, amylose is more concentrated at the peripheral parts of the granules than in their interior. However, in contrast, Blennow et al.[18] showed that amylose is more confined to the interior parts of granules from various starch plants.

Another intriguing question is to what extent amylose and amylopectin are associated with each other in the granule.[1] When starch granules are heated in a water suspension, they swell to a different degree depending on the temperature, and amylose tends to leach out from the granules. Amylose leaches more readily from maize starch granules than from potato starch granules. It was concluded that in potato, amylose is associated with amylopectin, but it is not associated, or associated to a lesser degree, in maize starch granules, as well as other cereals. In contrast, Jane et al.[19] did crosslink amylose with amylopectin in the granules and concluded that amylose is interspersed among amylopectin molecules in both maize and potato granules.

In summary, individual amylose chains are believed to be randomly located in a radial fashion among the amylopectin molecules.[14,20] Many experiments show that there is an enrichment of amylose toward the periphery of the granule and that the amylose found near the surface of the granule has a smaller chain length than amylose located nearer the center of the granule.[21] It is believed that amylose would fill spaces in the semi-crystalline matrix formed by amylopectin, probably rendering the starch granule denser.[2]

2.1.1.3 Helical Conformation

In a freshly prepared aqueous solution, amylose is present as a random coil.[22] The random coil conformation, however, is not stable. Amylose tends to form either single helical (inclusion) complexes with suitable complexing agents or double helices among themselves in the absence of a suitable complexing agent.

Several amylose crystalline forms (A, B, and C) can be found. The structure of the A-starch double helix in amylose is shown in Figure 2.2.[6,22] The crystalline forms only differ by the packing of the helices. The amylose double helices readily precipitate.[1] These precipitates are crystallites of the B-polymorph type described above.[9] The double helices in both A- and B-type are left-handed and parallel stranded.[6] They are almost perfectly a six-fold structure, with a crystallographically repeating unit of 10.5 Å. The symmetry of the double helices differs slightly in the A- and

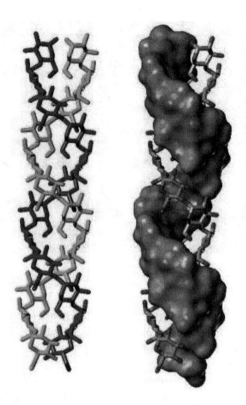

FIGURE 2.2   Three-dimensional structure of A-type starch double helix. (Reproduced with permission from Serge PEREZ.)[6]

B-structure. In the A-form, the repeated unit is a maltotriosyl unit, whereas in the B-type, the repeated unit is a maltosyl unit.

Single amylose molecules also form helices, which readily interact with a range of different compounds, such as iodine, fatty acids, or different alcohols to form inclusion complexes.[1] These left-handed helices are more compact than the double helices and one turn of the helix may comprise between six and eight glucose units depending on the guest molecule. Single amylose helices crystallize into the so-called *V-polymorph* pattern.[23–25] The single-helical (inclusion) complex was first reported by Katz[26] in the 1930s with cooked dough.[16] After cooking, the dough develops a diffraction pattern, which differs from the original A, B, or C pattern. Katz named it the V pattern from the German word "Vaklinestorone" (gelatinization). The V-amylose complexes are single, left-handed helices that are arranged as crystalline and amorphous lamellae, which may form distinct nanoscale or micron-scale structures (Figure 2.3).[27]

Direct evidence for the presence of amylose complexes with *lysophospholipids* in native cereal starches has been provided by ^{13}C cross-polarization/magic-angle spinning NMR.[1] Therefore, in cereals, the amylose is divided into lipid-complexed amylose (LAM) and free amylose (FAM). Observation of the amylose content of normal cereal starches (i.e., around 25%–30%) in comparison with the total lipid content (maximum ~1%) suggests, however, that only a small proportion of the amylose in

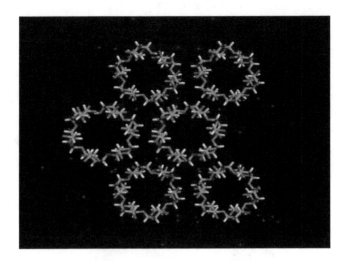

FIGURE 2.3   Structure of amylose-V repeat unit. (Reproduced with permission from Serge Perez.)[6]

such starches is complexed with lipid.[11] The well-known blue coloration of starch with iodine is due to the amylose–iodine complex and is much used to measure amylose in starch. However, LAM interferes with the measurement (in particular in cereal starches) and therefore only FAM is detected.

The transition from the coil to a single or double helix is attributed to the chemical structure of the amylose monomer, i.e., the 1,4-linked α-D-*glucopyranosyl* unit.[16] In a single-helical complex, the linear portion of the starch molecule has its hydrophobic side facing the cavity of the helix and interacting with the non-polar moiety of the complexing agent, such as the hydrocarbon chains of 1-butanol or fatty acids. With a complexing agent available in an aqueous solution, the single-helical complex forms instantaneously. In the absence of complexing agents, the linear portion of starch molecules will pair up to have their hydrophobic sides folded inside the double helix. The formation of a double helix requires an alignment of two molecules. Both single and double helices result in lower energy states and are thermodynamically favorable.

In 1985, Ring et al.[28] demonstrated that the majority of the amylose in the granule could be leached out of granules at temperatures just below the gelatinization temperature.[11] They further showed that most of the leached amylose chains were in the single helical state, rather than the double-helical state. Therefore, the single helical state is believed to be the predominant state of amylose chains within native starch granules. However, because amylose can only be completely extracted from granules at temperatures above 90°C, some large amylose molecules may be present within the granules and may participate in double helices with amylopectin.

In summary, amylose chains are believed to be in a single helical state, although a small proportion may be involved in lipid complexes.[21] The concentration of amylose (and lipid) increases toward the surface of the granule, with smaller (leachable) amylose chains predominating near the surface. Some of the larger amylose (non-leachable) chains may be involved in double-helical interactions with amylopectin.[14,15]

2.1.2 AMYLOPECTIN

Amylopectin is the major, highly branched component of starch.[1] The size of amylopectin is much larger than amylose. Amylopectin accounts for 75%–90% of wild-type starches, has a degree of polymerization (DP) of ~10^5, and has a branching level of 4%–5% (i.e., 4%–5% of its linkages are α-1,6 branch points).[2,29] Amylopectin forms the structural framework and underlies the semi-crystalline nature of starch.[2] Amylopectin has one reducing end and many non-reducing ends due to the branches occurring at every 12–30 residues along a chain of α-linked glucose units.[30]

The use of different enzymes to investigate the amylopectin structure is shown in Figure 2.4.[23,31]

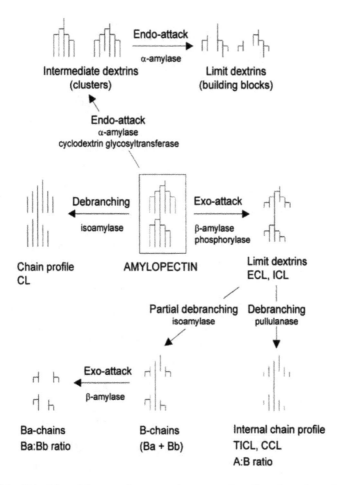

FIGURE 2.4 Principles of the use of enzymes in structural studies of amylopectin and other branched starch components. Amylopectin is drawn following the cluster model (see Section 2.1.2.4); α-amylase and cyclodextrin glycosyltransferase: endo-attack; β-amylase: exo-attack; isoamylase and pullulanase: debranching; CL, chain length; ECL, external chain length; ICL, internal chain length; A-chain and B-chain, outer and internal chains; TICL, total internal chain length; CCL, core chain length (see Section 2.1.2.1). (Reproduced with permission from Elsevier.)[29,31]

2.1.2.1 Chain Categories

2.1.2.1.1 Short and Long Chains

The chains of amylopectin are divided into two major groups, short (S) and long (L) chains; the division between the groups is generally at DP ~ 36.[1] Short chains predominate and the molar ratio of S:L chains varies between starches (Table 2.3).[1,4,32,33]

The S-chains form double helices that are involved in crystal formation in the starch granules. S-chains are anchored on the L-chains, which thereby function as the interconnecting chains and are mainly confined to the amorphous lamellae.

2.1.2.1.2 A-, B-, and C-Chains

In 1952, Peat et al.[34] suggested a useful nomenclature of the chains in amylopectin. A-chains are unsubstituted, whereas B-chains are substituted with other chains, and the C-chain carries the single reducing-end group of the macromolecule but is otherwise similar to the B-chains (Figure 2.5 and Table 2.3).[1]

In the models of amylopectin, the branch points are concentrated in certain regions from which the linear chain segments extend to form clusters, i.e., groups.[2] The frequency distribution of chain lengths deduced from the analysis of debranched starches shows that the chain lengths of most chains are between 10 and 20 glucose units (Figure 2.6).[35] These are considered to be the A- and B1-chains (B-chains that participate in the formation of only one cluster, also called BS for short B-chains). There are also longer chains (BL) that form connections between different clusters. The latter are generally believed to be oriented in the same orientation as the A- and B1-chains (cluster model) and assigned as B2- (DP ~42–48), B3- (DP ~69–75), and B4- chains for chains spanning two, three, and four clusters, respectively. They could, however, also be oriented perpendicularly to the clusters (the backbone model).[2] The C-chain has generally a broad size-distribution between DP 15 and 120, with a peak around DP 40.

In addition to S- and L-chains, several amylopectin molecules have also been found to contain extra-long chains (EL-chains), sometimes called super-long chains. EL-chains consist of several hundred or even thousand glycosyl units and thereby they resemble amylose chains and are difficult to distinguish from them.

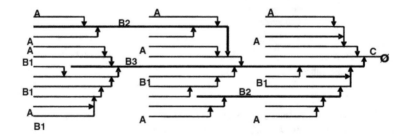

FIGURE 2.5 Structure of amylopectin, showing the A, B, and C chains. Adapted from Hizukuri (1986, 1996). Solid line: α-1,4-bound glucose units; arrows: α-1,6 linkage; φ: reducing glucosyl residue. (Reproduced with permission from Springer Nature.)[36]

(a)

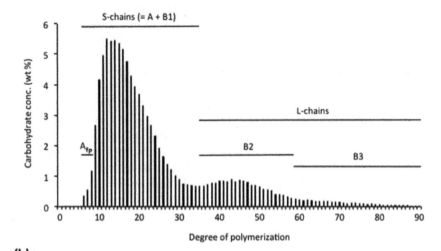

(b)

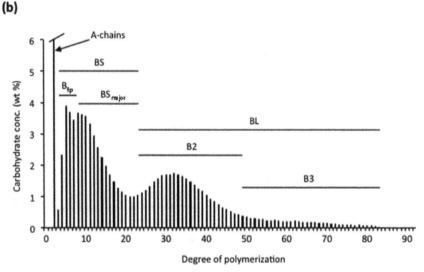

FIGURE 2.6 Unit chain profile of rice amylopectin (a) and its internal chain profile (B chains excluding external parts) (b). The A-chains (maltose peak) in (b) are actually external chain stubs. A-chains carry no branches; B-chains carry one or more branches; S-chains include A-chains and B1 chains; B2 and B3 are chains spanning two and three clusters, respectively. (Open access article distributed under the Creative Commons Attribution License.)[1]

2.1.2.1.3 External and Internal Chains

The unit chains of amylopectin contain external and internal chain segments (Table 2.3). External chains are defined as segments extending from the outermost branch point of a chain to its non-reducing end.[1,31] Thus, all A-chains are external, whereas only one part of the B-chain is external (Figure 2.7). The rest of the B-chain is called the total internal chain and includes all segments between the branches as well as all the glucose residues involved in the branch points. An alternative definition is the core chain, from which the outermost branch residue is excluded. Finally, internal chains are defined as the segments of the B-chains between the branches, excluding the branch point residues. The segment at the reducing end side is also considered an internal chain.[29]

Chain lengths and molar chain ratios in various amylopectins are shown in Table 2.3.

It is necessary to remove the external chains to study the internal chain structure of amylopectin (Figure 2.7).[1,38] Two amylolytic enzymes can be used for this purpose. Phosphorylase *a* removes successively one glucose residue from the non-reducing end through a phosphorolytic mechanism producing glucose 1-phosphate, whereas β-amylase produces β-maltose by hydrolysis. The enzymes were found to be exo-acting enzymes, i.e., they hydrolyze the chains from the non-reducing end until they approach the most exterior branch point, which possesses

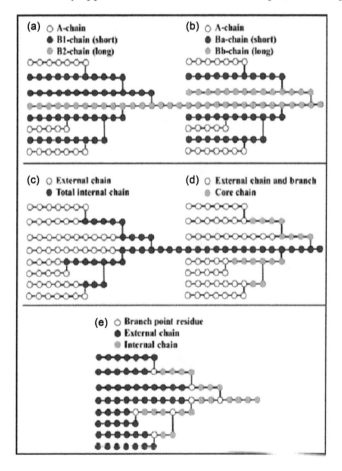

FIGURE 2.7 Structures of amylopectin. Dots represent glycosyl residues; horizontal lines represent α-1,4-glycosidic bonds; vertical lines represent α-1,6-glycosidic bonds. (Reproduced with permission from SAGE.)[37]

TABLE 2.3
Chain Lengths and Molar Chain Ratios in Amylopectins[1,a]

Source	CL	ECL	TICL	ICL	S:L	BS:BL	A:B
Wheat	17.7	12.3	12.7	4.14	16.2	6.8	1.4
Maize	19.7	13.1	12.6	5.6	9.9	6.3	1.1–1.2
Asian rice[b]	16.9	10.7	12.4	5.2	14.2	5.4	1.0
African rice[c]	18.1	12.1	11.9	5.0	11.7	5.0	1.0
Oat	17.0	10.7	12.6	5.3	18.2	8.6	1.0
Barley	17.6	11.5	12.3	5.1	17.9	8.6	1.0
Rye	17.4	10.7	12.6	5.3	16.2	7.3	1.1
Potato	23.1	14.1	19.9	8.0	6.3	2.3	1.2–1.5
Cassava	18.8	12.4	14.6	5.3	11.0	4.6	1.3

[a] CL, number average chain length, i.e., total number of glucosyl residues/number of chains; ECL, external chain length; TICL, total internal chain length; ICL, internal chain length; S:L, ratio of short to long chains; BS:BL, ratio of short to long B-chains; A:B, ratio of A- to B-chains.
[b] *Oryza sativa.*
[c] *Oryza glaberrima.*

a barrier the enzymes cannot bypass. The resulting resistant molecule is called a *limit dextrin* (LD) and contains the entire internal part of the original amylopectin together with shorter external chain stubs that the enzymes leave in front of the outermost branch points. Phosphorylase (φ) *a* produces a φlimit dextrin (φ-LD), in which all A-chains have been reduced into four residues (maltotetraosyl stubs). If the φ-LD is further hydrolyzed with β-amylase, each chain is reduced by an additional maltose residue, thus leaving the A-chains as maltosyl stubs in the so-called φ,β-LD. In the β-LD produced by β-amylase, the length of external residues depends on the odd or even number of glucosyl units in the original external segment. Therefore, A-chains remain maltosyl (DP2) or maltotriosyl stubs (DP3).

The internal chains are analyzed after debranching with isoamylase and/or pullulanase (Figure 2.6b).[1] It should be noted that there are some differences in the preferences for different substrates by the two enzymes: pullulanase more effectively hydrolyses maltosyl chain stubs, whereas isoamylase more effectively hydrolyses the whole amylopectin. Glucosyl branch stubs are resistant to both enzymes. The ratio of A:B-chains is normally between 1.0 and 1.4 (Table 2.3). The internal BSs are distinguished as two subgroups: because the size-distribution profile of the shortest chains with DP 3–7 (and the peak at DP 5 or 6) appeared to be specific for a particular plant, the name "fingerprint" B-chains (B_{fp}) was suggested, whereas the major part of the short B-chains with DP 8–23 was called BS_{major}-chains.[38] The long internal B-chains with DP > 23 (BL-chains) correspond apparently to the same categories of B2- and B3-chains as found in the whole amylopectin. In the internal chain profile, they are shorter, however, as the external segment has been removed (DP 23–28). Moreover, the A-chains have been subdivided into two groups.[1] The shortest chains in the unit chain profile of the whole amylopectin at DP 6–8 are considered as A-chains and have been called "fingerprint" A_{fp}-chains (Figure 2.6a) in analogy to internal B_{fp}-chains as it was shown that chains with DP < 9 do not readily form double helices. The rest of the A-chains was considered to be crystalline A-chains ($A_{crystal}$).

The unit chain profile of the whole amylopectin can be used for the estimation of the average chain length (CL). By combining the information from both the unit and internal chain profiles, the average external, total internal, and internal chain lengths (ECL, TICL, and ICL, respectively) can also be calculated (Table 2.3).

2.1.2.2   Structural Types

The internal chain profile of amylopectin was found to be specific for different plant species and was divided into four structural types (Figure 2.8).[1]

Type 1 amylopectin, which includes A- and C-crystalline starches, is typical for certain cereals such as barley, rye, and oat. In these starches, the relative number of long chains is very small with hardly any B3-chains. The ratio of S:L-chains (as well as BS:BL) is therefore high. The BS-chains possess an unusually broad size-distribution because of a great amount of BS_{major} chains and therefore the groove between BS and BL-chains has almost completely disappeared.

Type 2 amylopectin, which also includes A- and C-crystalline starches, has more B-chains than type 1 and few BS_{major} chains, and therefore, the groove between BS- and BL-chains is clearly distinguished. In addition, B_{fp}-chains are more abundant than in most other structural types. The important cereals maize and rice belong to

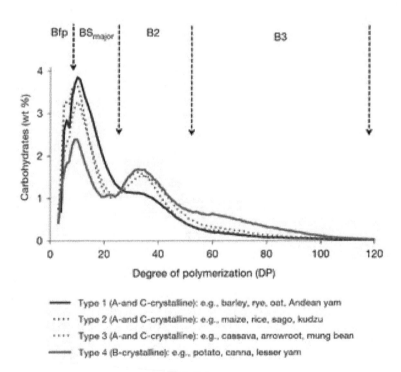

FIGURE 2.8   Internal unit chain profile of debranched φ-β-LDs from the four different types of amylopectin, as obtained by high-performance anion-exchange chromatography. (Adapted from Ref. 39. Reproduced with permission from Elsevier.)[63]

type 2 amylopectin, whereas wheat amylopectin was found to have an internal chain structure intermediate to that of types 1 and 2.

Type 3 amylopectin, which also includes A- and C-crystalline starches, is typical for cassava and mung bean starch and has fewer B_{fp}-chains than types 1 and 2 and slightly elevated B3-chains.

Type 4 amylopectin, which includes all B-crystalline starches (e.g., potato and edible canna), possesses the highest amount of B3-chains and thus the lowest S:L chain ratio.

2.1.2.3 Branched Units

The chains in amylopectin are arranged into larger or smaller groups and diverse groups can be isolated using endo-acting enzymes. The α-amylase of *Bacillus amyloliquefaciens* has been the mostly used enzyme for investigating branched structural parts of diverse amylopectins. Owing to its most pronounced endo-action pattern, the enzyme is especially useful for the isolation of larger branched α-dextrins from amylopectin (Figure 2.9). The reaction is effective only with all subsites of the enzyme filled with glucosyl units. Initially, the rate of the reaction is very fast because of the presence of several long internal chain segments that completely fill up the subsites.

As the α-amylase also attacks the external chains in amylopectin in a more uncontrolled fashion, the remaining external segments have diverse lengths.[1] The LDs of the clusters or domains can then be structurally characterized. The size-distribution of clusters is large, varying from approximately DP 15 up to DP 500 or more. The largest clusters are found in type 1 and 2 amylopectins, which include the cereals, having average size around DP 66–73. This corresponds to an average number of chains (NC) of about 11–14 (Table 2.4).

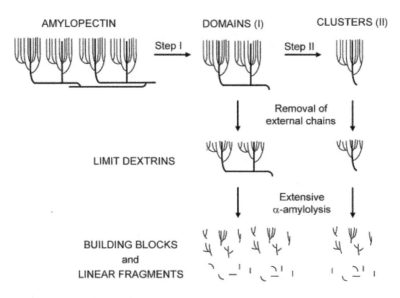

FIGURE 2.9 Amylopectin *building blocks* defined as the smallest branched units inside clusters. (Reproduced with permission from Elsevier.)[39]

The apparent clusters isolated from amylopectins are very slowly hydrolyzed further by the α-amylase into small α-LDs, which constitute the ultimate branched units called *building blocks* (Figures 2.10 and 2.11).

The molecular composition and the detailed structure of building blocks are very complex.[1] Nevertheless, the concept of building blocks can be rationalized under diverse groups of α-LDs with an increasing number of chains. Thus, group 2 consists of building blocks with two chains and has DP 5–9; group 3 has three chains

TABLE 2.4

Structure of Intermediate α-Dextrin "Clusters" Isolated from Amylopectin with α-Amylase[1,a]

Source	Structure[b]	NC	CL	NBbl	Molar Distribution of Building Blocks (%)[c]				
					2	**3**	**4**	**5**	**6**
Wheat	A:1–2	14.2	5.8	6.3	57	24	10	8	1
Rye	A:1	11.5	6.1	5.5	55	28	9	7	1
Barley	A:1	19.5	6.3	10.4	65	25	5	4	1
Oat	A:1	11.8	6.1	5.7	55	27	9	7	2
Asian rice[d]	A:2	12.0	6.8	5.7	50	29	11	9	1
African rice[e]	A:2	14.1	5.9	5.1	40	29	13	14	4
Maize	A:2	12.5	5.4	4.2	51	27	10	12[f]	
Cassava	A:3	10.0	7.0	5.1	59	30	5	6[f]	
Waxy potato	B:4	6.2	7.9	3.1	48	31	13	8[f]	

[a] NC, average number of chains in the cluster; CL, number average chain length, i.e., total number of glucosyl residues/number of chains; NBbl, average number of building blocks in the cluster.

[b] Type of crystalline structure and type of amylopectin molecular structure.

[c] Building blocks constitute the smallest branched units found inside clusters; building blocks of group 2 have two chains, group 3 have three chains, group 4 have four chains, group 5 have 5–7 chains, group 6 have >7 chains.

[d] *Oryza sativa*.

[e] *Oryza glaberrima*.

[f] Group 5 + group 6.

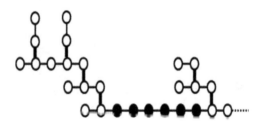

FIGURE 2.10 Glucose residues in amylopectin organized into building blocks (unfilled circles) and separated by residues in inter-block segments (black circles). (Reproduced with permission from Elsevier.)[40]

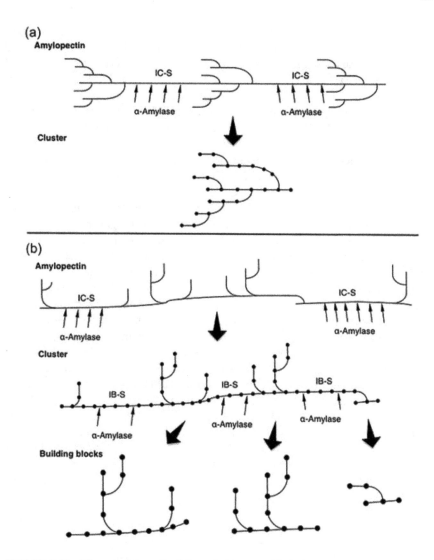

FIGURE 2.11 The principles of isolation of clusters and building blocks from amylopectin with α-amylase; (a) based on the cluster model and (b) based on the backbone model (see text). Only the internal parts of the molecules are shown to highlight the endo-attack of the enzyme. IB-S; inter-block segment; IC-S, inter-cluster segment. (Reproduced with permission from Elsevier.)[31]

(DP 10–14) and group 4 has four chains (DP 1519). Group 5 consists of dextrins with 5–7 chains and group 6 has ~10 chains and DP > 35. The chains in the building blocks are very short: besides a-chains (DP 2–3), most b-chains are concentrated around DP 5–6. In the larger building blocks, chains with DP 12–15 can be detected. Group 2 building blocks predominate in the clusters: roughly 40%–65% of all building blocks (on a number basis) have only two chains. The amount then decreases with an increasing number of chains, and the size-distribution is similar regardless of the structural type of amylopectin.

The largest building blocks in group 6 contains as many chains as found in a large part of the clusters.[1] However, the chains in the building blocks are much shorter and the density of branches is higher (14%–20%) than in clusters (11%–15%) because of very short distances between the branches; the ICL is only 1.4–2.3. The large clusters in cereals have typically 4–6 building blocks, whereas small clusters (many roots and tubers) contain 3–5 blocks on average (Table 2.4). The CL between the building blocks (the inter-block CL, IB-CL) increases successively in clusters from type 1 to type 4 amylopectins from IB-CL ~5.7 to 8.0. In only a few cases, the inter-cluster CL (IC-CL) has been estimated in isolated domains from amylopectins.[1] IC-CL appears mostly to be between 10 and 20. The shortest IC-CL (9.5) was estimated in sweet potato and barley and the longest was estimated (IC-CL 27) in a domain of waxy potato. IB-CL was shown to correlate positively with the onset of gelatinization temperature. The inter-block segments, and most probably also the inter-cluster segments, have probably a key role when it comes to contributions to the thermal properties of starch granules.

It is the b-chains in the clusters that interconnect the building blocks, i.e., they consist of inter-block segments as well as intra-block chain segments.[1] The b-chain is given a number that corresponds to the number of inter-block segments along the chain (Figure 2.12). Thus, b0-chains lack inter-block segments, because they are found completely inside the building block. The DP of these chains is 3–6 and, if not formed by the α-amylase, they correspond to B_{fp}-chains in the original amylopectin. b1-Chains contain one inter-block segment. They have DP 7–18 and are subdivided into b1a-chains (DP 7–10), which practically lack an internal segment at the reducing-end side of the chain, and b1b-chains (DP 11–18), which have a segment at the reducing-end side that extends through an internal building block. b2-Chains (DP 19–27) have two inter-block segments and segments that extend through building blocks. Finally, b3-chains (DP $\geq$ 28) have at least three inter-block segments. Roughly, one can consider that each building block on a b-chain contributes with an intra-block segment (DP 5–6) in addition to an inter-block segment (DP 5–8) that connects to the next block, i.e., the b-chains theoretically possess a periodicity of around 12 residues.

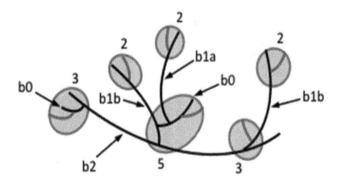

FIGURE 2.12 Chain types and building blocks in an isolated typical "cluster"; a-chains are shown (red); b-chains (black) are numbered based on the number of inter-block segments they are involved in. Building blocks are encircled (gray) and numbered according to the number of chains they contain. (Open access article distributed under the Creative Commons Attribution License.)[1]

2.1.2.4   Organization of Structural Units

The challenge in understanding the structure of amylopectin is to organize the information on the structural units into a model of the entire macromolecule.[1]

In this section, two models are described, namely the "cluster model" and the "building block backbone model". The former is the more traditional model, and the latter is the most recent model based on findings now available.

2.1.2.4.1   The Cluster Model

The cluster model was originally proposed in 1969 by Nikuni[41] and independently by French[42] in 1972. In this model, the short chains in amylopectin form clusters, and the long chains interconnect the clusters (Figure 2.13a). The model is based primarily on the following data. (1) Acid treatment of starch granules removes mostly the amorphous parts, whereas the crystalline parts remain largely intact. Size-distribution analyses of the molecules in the remaining granular residues show that most of the branches have been removed and the major part of the dextrins consists of linear chains with DP 13–16. The branches are probably mostly confined to the amorphous parts and connect to the short chains in the crystalline parts. (2) The periodic length of 9–10 nm corresponds to one amorphous and one crystalline lamella, in which the thickness of the latter closely matches the DP of the short chains in the acid-treated starch. (3) Amylopectin consists of two major groups of chains (S- and L-chains). (4) The molecules are arranged radially with all chains pointing in the same direction toward the surface of the granules.

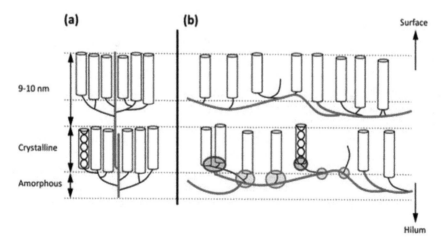

FIGURE 2.13   Amylopectin in the semi-crystalline growth ring depicted based on the cluster model (a) or the building block backbone model (see text) (b). Long B-chains and short chains are shown. The backbone in (b) carries internal building blocks (some encircled) and can also carry short BS$_{major}$-chains that form branches to the backbone and connect to external building blocks (encircled). Cylinders represent double helices formed by external chains. The directions toward the surface and the hilum of the granule are indicated at the right. (Open access article distributed under the Creative Commons Attribution License.)[1]

Additional work suggested that dextrins formed from amylopectin were the clusters in the molecule. Moreover, Hizukuri[43] found a periodicity in chain length of 27–28 glucosyl units among the B-chains of amylopectin, which closely corresponds to the periodic length of 9–10 nm, i.e., the length of a chain in the form of a double helix with six residues per turn and a pitch of 2.1 nm.

Several authors have challenged the cluster model. At first, Hanashiro et al.[44] found a periodicity in chain length of 12 glucose residues among the S-chains, but not among the L-chains. It appears that the periodicity of 12 residues stems from the interconnection of building blocks in the isolated α-dextrins. Secondly, Sullivan and Perez[45] showed that the amorphous chain segments involved in interconnecting double helices into parallel alignment take almost perpendicular directions to the helices. Thirdly, Bertoft and co-workers[46] found that the structure of some branched dextrins does not conform to the expected structure of clusters proposed in the cluster model. These findings led to a need for a new model.

2.1.2.4.2 The Building Block Backbone Model

The new model is called the building block backbone model.[47] In this two-directional backbone model, the clusters are connected to a backbone that extends in an almost perpendicular direction and is formed by the amorphous chains.[48] The model makes it possible to construct a super-helical structure from a single amylopectin macromolecule.

In the building block backbone model, the L-chains (or BL-chains) in amylopectin are linked to each other and form collectively a longer backbone (Figure 2.14b). Building blocks are outspread along the backbone and form an integrated part (these blocks are characterized as "internal"). They are completely randomly distributed along the backbone.[49] The short chains (S-chains) extend from the building blocks in a more or less perpendicular direction. In the starch granules, the S-chains form the double helices of the crystalline lamellae, i.e., exactly as in the cluster model. In contrast to the cluster model, however, the backbone is completely embedded in the amorphous lamella and extends along it, rather than traversing the stacks of lamellae in the semi-crystalline "growth ring". The building blocks are separated by inter-block segments (DP 5–8; see Table 2.4 and Figure 2.13) and occasionally, these segments are slightly longer, forming the inter-cluster segments. It is apparent, however, that the actual function of the inter-cluster segments is the same as that of the inter-block segments; they interconnect building blocks. Therefore, the branched structural units in the model are the building blocks.

The α-dextrins that the α-amylase produces at the early stages of hydrolysis are therefore only apparent clusters, which are formed because of the preferential attack at inter-block segments (DP ≥ 9). This explains why the structure of the α-dextrins does not correspond to the cluster model. Instead, the backbone model explains the fact that some dextrins still contain long chains. The long chains are involved in interconnecting building blocks and, depending on the random placement of long inter-block segments, the released dextrins will contain a variable number of building blocks, inter-block segments, and chain lengths, some of which correspond to the long b3-chains.[50] This also explains the very broad size-distribution of the α-dextrins. The length of both the intra- and inter-block segments along the backbone varies randomly and, therefore, there is no apparent periodicity in chain length among the

L-chains.[51] Among the shorter chains (b2- and b1-chains in α-dextrins), there is a periodicity, however, because these chains are mostly found as short side-chains to the backbone and have well defined inter-block chain length.[1]

2.2 THE STARCH GRANULE

In view of the wide variety of granular morphologies, it is surprising to find that their architectural features are shared universally among plants and plant organs.[1] Most of the native starch granules exhibit a Maltese cross when observed under polarized light (see Figure 5.1 in Chapter 5).

The radial organization of the amylopectin within such structures is thought to cause optical polarization because the visible optical polarization is in the order of the wavelength of the visible light (100–1,000 nm) (S. Perez, Personal communication, 2020). Native granules yield X-ray diffraction patterns of low quality, as the bulk of the starch polymers are in an amorphous state (70% on average).

Starch granules from different species and tissues vary greatly in size and shape, ranging from relatively small particles of 0.5–2 μm in diameter in amaranth seeds and flat disks in *Arabidopsis* leaves to smooth spheres of up to 100 μm in tuberous roots.[2] They are made up of alternating amorphous and semi-crystalline shells which are between 100 and 400 nm thickness.[6] These structures are called *"growth rings"*. The radial organization of the amylopectin within such structures causes optical polarization.

Granules contain small amounts of protein (typically 0.1–0.7%), which is mostly the granule-bound starch synthase that makes amylose, and also other amylopectin synthesizing enzymes, such as other starch synthases and starch-branching enzymes.[2] Many starches further contain traces of lipids and phosphate groups (covalently linked at the C6 or C3 position of glucose). The phosphorylation level of cereal starches is extremely low. In *Arabidopsis* leaf starch, it is around 0.05%, whereas in tuber starches, it can be many times higher (~0.5% in potato). Phosphorylation appears to be confined to amylopectin and enriched in the amorphous regions. A high phosphate content is an industrially relevant trait as it is associated with increased granule hydration and lowered crystallinity, yielding starch pastes with higher transparency, viscosity, and freeze–thaw stability.

Blocklets constitute the units of the growth ring.[52] They are made of amorphous and crystalline *lamellae* (9–10 nm) containing amylose and amylopectin. X-ray diffraction investigations indicate a periodicity of 9–10 nm within the granule. The periodicity is interpreted as being due to crystalline and amorphous lamellae found within the semi-crystalline shells. Perez et al.[48] demonstrated that the architecture of the blocklets follows phyllotactic rules. Amylopectin double helices form the crystalline lamellae of the blocklets. Starch nanocrystals result from crystalline lamellae when separated by acid hydrolysis. The levels of organization of the starch granule are shown in Figure 2.14.[49]

Figure 2.15 shows the amylopectin molecule, the organization of the molecules in crystalline and amorphous regions, and their radial orientation in the granule.[5]

Each amylopectin molecule contains up to two million residues in a compact structure.[5] The molecules are oriented radially in the starch granule. As the radius

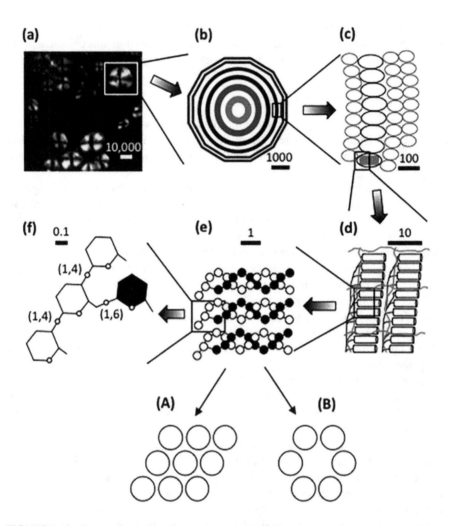

FIGURE 2.14 Starch: from granules to glucosyl units.[1] The bar scale (in nm) is only approximate to give an impression of the size dimensions. (a) Maize starch granules observed under polarized light showing the "Maltese cross", which is indicative of a radial organization within the starch granule. (b) A hypothetical granule (in this case polyhedral) with "growth rings" extending from the hilum. (c) *Blocklets* in semi-crystalline (black) and amorphous (gray) rings.[55] (d) Crystalline and amorphous lamellae formed by double helices (cylinders) and branched segments of amylopectin (black lines), respectively. Amylose molecules (red lines) are interspersed among the amylopectin molecules. (e) Three double helices of amylopectin. Each double helix consists of two polyglucosyl chains, in which the glucosyl residues are symbolized by white and black circles, respectively. The double helices form either A- or B-polymorphic crystals (A and B, respectively, in which the circles symbolize the double helices seen from the edge). (f) Glucosyl units showing α-1,4- and α-1,6-linkages at the base of the double helix. (Open Access article distributed under the Creative Commons Attribution License.)

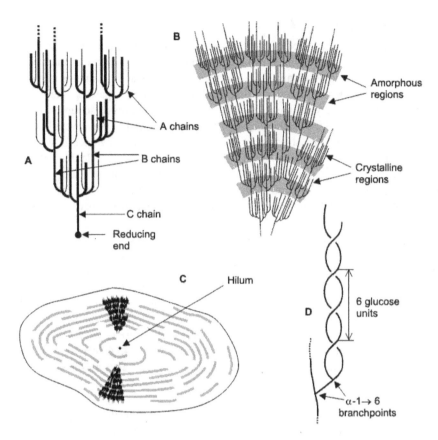

FIGURE 2.15 **Structure of a starch granule.**[5] (a) The essential features of amylopectin (A-chains: outer unbranched chains; B-chains: inner branched chains; C-chain containing the single reducing group). (b) The organization of the amorphous and crystalline regions of the structure generating the concentric layers that contribute to the growth rings. (c) The orientation of the amylopectin molecules in a cross-section of an idealized granule. (d) The likely double helix forming type A and type B amylopectin crystallites. It appears most likely that the crystalline structure consists of parallel left-handed helices with six residues per turn. (Creative Commons Attribution).

increases, so does the number of branches required to fill up the space, with the consequent formation of concentric regions of alternating amorphous and crystalline structure. The structure of starch granule is depicted from the amylopectin cluster, to the organization of amorphous and crystalline regions, to the orientation of the amylopectin molecules, and finally to the double helix structure formed between two neighboring chains. Amylopectin double-helical chains can form either the more open hydrated type B hexagonal crystallites or the denser type A crystallites, with staggered monoclinic packing, dependent on the plant source of the granules.

2.2.1   Crystallinity

As already mentioned, starch consists of two major molecular components: amylose and amylopectin.[56] Some plants also contain materials with structures intermediate to amylose and amylopectin. As described before, starch granules are semi-crystalline.[1] If they are treated in dilute acid, the amorphous parts in the granules are removed and the crystalline parts remain. The Maltese cross also remains, which shows that the organized molecular segments are confined to the crystallites. The crystallites are formed by short, external chain segments of amylopectin with a DP of approximately 10–20 glucosyl units. As two chains join into a double helix with six glucose residues per turn of each strand and a pitch of 2.1 nm, the length of these double helices is about 4–6 nm. The double helices crystallize into the so-called A- or B-type crystals. Some plants possess granules with a mixed pattern, termed as the C-type.

A-type starches can be found in cereal endosperms, whereas B and C types were reported in tubers and pea embryos, respectively.[57] Three-dimensional structures involving parallel-stranded double helices have been published for both the A and B patterns (Figure 2.16). Both diffraction patterns are due to an ordered array of double helices. The double helices are stabilized by a network of hydrogen bonds.[58] They are identical for both A and B-types but their packing manner and water content differ.[1] In the A-type crystal, the double helices are closely packed into a monoclinic unit cell (with dimensions a = 20.83 Å, b = 11.45 Å, c = 10.58 Å, space group B2) containing eight water molecules. In the B-type crystal, the double helices are packed into a hexagonal unit cell (dimensions a = b = 18.5 Å, c = 10.4 Å, space group P6$_1$) containing 36 water molecules. In this crystalline lattice, the water molecules fill up a channel, which does not exist in the A-type. The relative crystallinity in starch granules greatly varies between plant varieties in the range of 17%–50%, most often being higher in waxy starches compared to their amylose-containing counterparts. The double helices are stabilized by a network of hydrogen bonds.

The three-dimensional structures of these crystals are of paramount importance in determining most of the functional and industrial properties of starches.

2.2.1.1   A-Type Crystals

In 2006 and later in 2009, Popov et al.[59,60] determined the three-dimensional structure of A-amylose crystals as a model of the crystal domains of A-starch granules from synchrotron radiation microdiffraction (Figures 2.17 and 2.18).

The structure of A-type crystals is based on the repeat of a *maltotriosyl* residue, and there are four such residues per unit cell. The labeling of the atoms along one chain is shown in Figure 2.19.[61]

The structure indicates a "parallel-down" organization of the amylose molecules within the unit cell, indicating that the non-reducing end of the amylose molecules is oriented toward the positive direction of the unit cell *c* axis. The description of this geometry is important to correlate the crystallography of the granules of A-starch with their ultrastructure and their mode of biosynthesis. Indeed, it is known that starch synthases elongate starch molecules by adding D-glucopyranosyl units at the

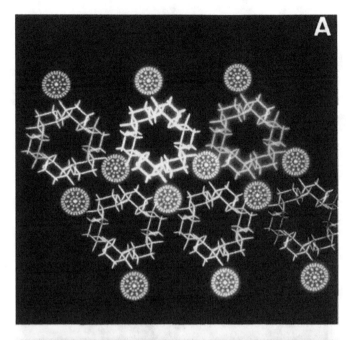

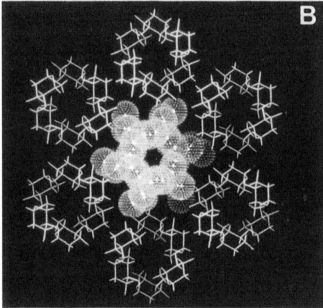

FIGURE 2.16 Three-dimensional representation of amylopectin crystalline structures. (a) Cross-section of the A-type lattice of parallel-stranded glucan double helices. The position of water molecules is displayed as orange spheres. (b) Cross-section of the B-type lattice of parallel-stranded glucan double helices. Water molecules are displayed in blue. (Reproduced with permission from the American Society of Plant Biologists.)[57]

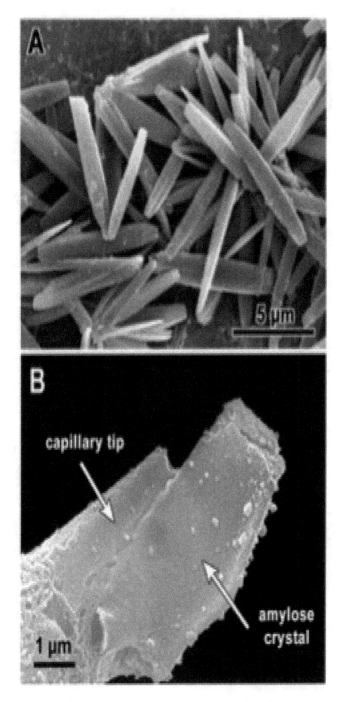

FIGURE 2.17   (a) Scanning electron microscopy (SEM) image of A-type amylose crystals prepared by the recrystallization of synthetic amylose. (b) SEM image of an amylose single crystal (arrow) glued to a borosilicate glass capillarity tip (arrow). The small ripples on the glass capillarity are due to the glue. (Reproduced with permission from ACS Publications.)[60]

FIGURE 2.18 Projection on the (a, b) plane of the structure of A-type crystals ($a = 20.83\,\text{Å}$, $b = 11.45\,\text{Å}$).[62] In such a unit cell, 12 glucopyranose units are located in two left-handed, parallel-stranded double helices, packed in a parallel fashion. For each unit cell, water molecules (red circles) are located between the helices. (Reproduced with permission from ACS Publications.)

non-reducing ends of amylopectin, therefore also toward the positive direction of the c-axis of the A-amylose lattice. In a given amylopectin molecule, there is only one reducing end but multiple non-reducing ends (as many as branches) are located at each pendent end of the short amylose fragments that branch out of amylopectin and form the A-type crystals. The fact that starch granules grow by apposition, because newly formed starch is deposited at or near the granule surface, would indicate that in the radial organization of the amylopectin chains, the reducing ends of these molecules should point toward the center of the granule, with the consequence of a radial organization of the positive direction of the c-axis, oriented from the granule center toward its periphery.

2.2.1.2 B-Type Crystals

In 1988, Imberty and Perez[61] established the structure of the B-type crystals from X-ray fiber and electron diffraction (Figures 2.16b and 2.20).

In 2004, Takahashi et al.[62] prepared a fiber specimen of B-starch from enzymatically synthesized amylose, which is a purely linear 1,4-linked poly-α-D-glucose. The conformation of the primary hydroxyl group is gg, and the angles of the bonds linking two glucose residues, Φ and Ψ, are $+84.3°$ and $-135.2°$, respectively.

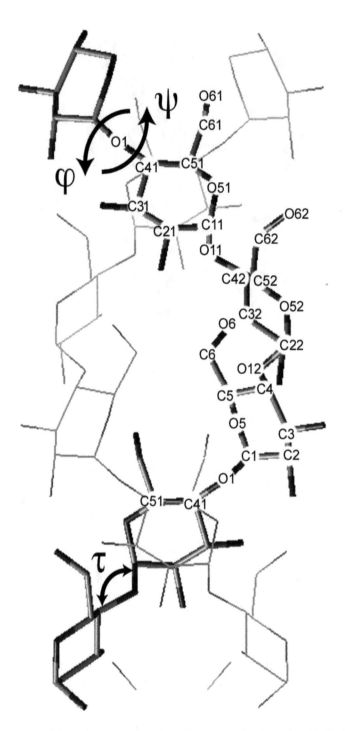

FIGURE 2.19 Labeling of the carbon and oxygen atoms together with the torsion angles φ and ψ and the bond angle τ along one of the strands of an A-type double helix.[72] The double helices are identical for both A- and B-types. (Reproduced with permission from ACS Publications.)[58]

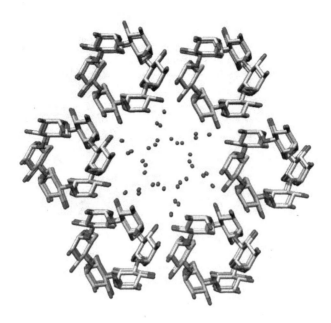

FIGURE 2.20 B-type starch; for each unit cell, water molecules (circles) are located between the helices. (Reproduced with the permission of Serge Perez.)

2.2.1.3   C-Type Crystals

The micro-focus synchrotron wide-angle diffraction mapping experiments were also used to decipher the crystalline microstructure and the polymorphism of granules from smooth pea, exhibiting the C-type polymorphism (S. Perez, Personal communication, 2020).[63,64] These granules contain 60% of A-type structure and 40% of B-type structure, and these two crystalline phases co-exist within the same granule. The A-allomorph is essentially located in the outer part of the granules, whereas the B-type is found mostly in their center. In the A-component, the diffraction diagrams were always poorly oriented fiber patterns with their fiber axis systematically oriented toward the center of the granule. In the B-type center of the granule, only powder diagrams could be observed, whereas in the middle areas of the granules, the B-component was much better oriented than the A-component. These observations bring a definite answer to the nature of the C-allomorph that is a combination of the A- and B-crystalline constituents. The synchrotron micro-beam used for this investigation had a diameter of around 2 μm, which is small enough to assess the gross ultra-structural features of the diffracting materials of individual starch granules. However, this beam size was still too large to assess the details of the crystalline micro-morphology of the individual concentric layers, of about 500 nm in thickness, that are located within the granule.

2.2.2   Lamellar Structure

Small-angle X-ray scattering shows a 9–10 nm repeat distance in all starch granules.[1] This was suggested to stem from stacks of repeating crystalline and amorphous lamellae, of which the former are represented by the double helices. The crystalline

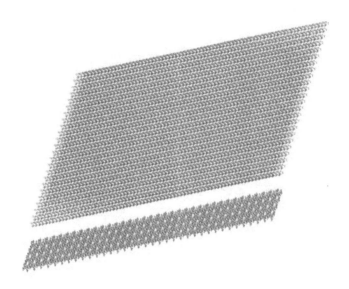

FIGURE 2.21   A starch nanocrystal. (Reproduced with the permission of Serge Perez.)

lamellae have been isolated from acid-treated granules. Waxy maize granules with A-type crystallinity show regularly formed parallelepipedal blocks having acute angles of 60°–65° and with lengths and widths of 20–40 and 15–30 nm, respectively. The dimensions suggest that a single nanocrystal contains between 150 and 300 double helices (Figure 2.21).

Amylose-containing starches give rise to nanocrystals with less perfect symmetry, suggesting an interference of amylose on the crystal structure. Nanocrystals isolated from B-type granules possess irregular structures, probably a result of the organization of the double helices in the B-crystallites. As the repeat distance is 9–10 nm and the crystalline lamellae are 4–6 nm, the amorphous lamellae are 3–6 nm thick. This part consists of longer, internal chains of amylopectin and probably amylose chains.

2.2.3 GROWTH RINGS

The stacks of amorphous and crystalline lamellae form rings, being in the order of 100–400 nm thickness.[1] The rings are semi-crystalline in nature as they contain amorphous and crystalline lamellae, but they have also been considered as "crystalline" or "hard" shells. The rings are embedded in an amorphous matrix with a lower molecular density in potato (0.49 g/cm^3) than in waxy maize (0.68 g/cm^3) and other A-crystalline starches. The matrix was described as an "amorphous background", "semi-crystalline shell", or "soft shell". The growth rings are generally thinner at the periphery of the granules and thicker in their interior parts. The relative crystallinity decreased somewhat in granules from constant light exposure. Around the hilum of the granules, the rings might be absent and replaced by a void (filled with water, if the granule has not been dried). In some starch granules, especially in many cereals (e.g., maize, wheat, barley, and sorghum), channels penetrate the granules from the surface. The channels are important sites of penetration of different enzymes, such as α-amylases or glucoamylases, and diverse chemicals.

2.2.4 BLOCKLETS

2.2.4.1 Structure

Scanning electron microscopy (SEM) and atomic force microscopy have revealed protrusions, named blocklets, at the surface of the granules with sizes ranging from 10 to 300 nm (average size: 100 nm).[1] The size appears to be partly species specific. Potato granules have larger blocklets (50–300 nm) than wheat granules (10–60 nm). They are proposed to contain 280 amylopectin side-chain clusters.[65] Blocklets have also been observed within the growth rings in the interior parts of starch granules and appear to transverse the entire rings. SEM images of residual starch granules reveal that wheat starch has smaller blocklets with a diameter of ~25 nm in the soft amorphous shells and larger blocklets with a diameter of 80–120 nm in the hard crystalline shells.[32]

2.2.4.2 Phyllotactic Rules Applied to the Architecture of Blocklets

Starches display a series of structural features that are the fingerprints of levels of organization over six orders of magnitude.

Perez et al.[53] showed that general rules control the morphogenesis of the hierarchical structures. They considered the occurrence of phyllotaxis-like features that would take place at scales ranging from nano- to micrometers. Based on the Fibonacci golden angle, they derived theoretical models to simulate X-ray and neutron scattering patterns. Among these models, a golden spiral ellipsoid displayed shapes, sizes, and high compactness reminiscent of the blocklet.

From the convergence between the experimental findings and the theoretical construction, they suggested that the phyllotactic model represents an amylopectin molecule, with a molecular weight in the order of 10^8–10^9 Da. In this model, the amylopectin molecule would be composed of a series of crystalline platelets and amorphous layers having a width of 9 nm and being made of parallel-stranded double helices. The amylopectin constituting the blocklet would have the form of an ellipsoid (a "cigar" shape) that has been experimentally observed (Perez-Herrera et al., 2017).[66] Figure 2.22 illustrates the principle of construction and the results obtained for a typical ellipsoid having a height of 200 nm with lateral dimensions of 12 nm and 74.5 nm, respectively. However, a variety of possible structures having similar shapes can be formed. The model, referred to as phyllotactic, displays a high level of compactness of the constituting elements and integrates all the key three-dimensional data reported so far.

2.3 CONCLUSIONS

As early as 1858, Carl von Nägeli[67] stated, "The starch grain opens the door to the establishment of a new discipline: the molecular mechanics of organized bodies".[15] Thus, despite the progress made in understanding starch granule structure, only part of this Pandora's Box has been opened, although we are beginning to explore its boundaries. The information gathered so far lets us recognize the extreme complexity of the starch granule as a macromolecular assembly. While the whole picture of the starch granule architecture appears better understood, much of the progress reached consists of pieces of information regarding specific features of the granule architecture. A consistent description of the relationship between these different levels of structural organization is still lacking. It is certainly useful to distinguish between those features that have received widespread acceptance and those that require further investigation.

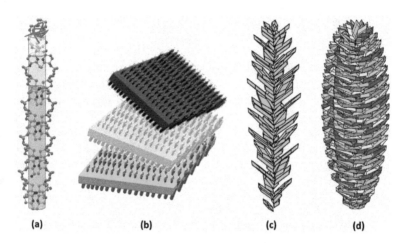

(a) **(b)** **(c)** **(d)**

FIGURE 2.22 Starch: from double helix to blocklet. (a) From the crystal structure of A-type starch and the results of computational modeling, a prototypical motif can be constructed that shows how the left-handed parallel-stranded double helices are connected by an α-1,6-linkage. The cumulative distance of these structural arrangements is about 6.5 nm of crystalline order that is followed by a 2.5 nm of the less-ordered constituents. This accounts for the 9 nm repeat recurrently observed in all starches. (b) Schematic representation of the relative spatial organization of three neighboring platelets, which are stacked and rotated with respect to each other, according to the Fibonacci golden angle. The dimensions and content of each starch platelet agree with the experimentally observed features of nanocrystals that have 5–7 nm thickness, characteristic geometrical features such as 60°–65° acute angles, and form parallelepipedal blocks that contain several hundreds of double helices. To these, a layer of 2–2.5 nm is added to model the embedded less-ordered constituents. The color-coding refers to the spiral shown in (c) to which each platelet belongs to. (c) Representation of one golden spiral built from the progressive spatial organization of platelets described in (b). The number of layers is 31. The color-coding of the spiral (green) refers to the color-coding of the green platelet shown in (b). (d) Representation of one golden spiral ellipsoid as a model of starch blocklet. The ellipsoid is made up of seven helices as the one shown in (c). The platelets of each of the seven spirals are represented with the same color (red, orange, yellow, green, cyan, blue, and violet). The height of the golden spiral ellipsoid is 200 nm, with lateral dimensions of 12 nm and 74.5 nm, respectively. (Reproduced with the permission of Serge Perez.)

Although the size and morphology of the starch granules are specific for each plant species, their internal structures have remarkably similar architecture, consisting of growth rings, blocklets, and crystalline and amorphous lamellae, down to the double-helical arrangement of the short, external chains of amylopectin. The structural levels span a size-range from micrometers (granular level) to nanometers (molecular level).[1,56] The branched, major component amylopectin is generally agreed to be the principal contributor to the semi-crystalline structure of the granules.

The following are established features derived from recent applications of instrumentation and methods to the three-dimensional elucidation of complex bio-macromolecules[15]:

- The occurrence of segments of amylopectin with a left-handed, parallel-stranded, double-helical structure.

- Two types of stable arrangements of double helices can occur, as found in the A and B polymorphs.
- Up to 200 double-helical segments are densely packed in platelet nanocrystals, which are polar and chiral.
- The 1–6 branching does not preclude the formation of double-helical arrangements.
- The amorphous amylopectin may contain a significant amount of double helices.

As opposed to the more traditional cluster model, the building block backbone model of amylopectin is based on new findings obtained on α-dextrins. In addition, the building block backbone model is compatible with the former data in favor of the cluster model and it explains satisfactorily many of the properties of starch granules. Moreover, it implies new ways to look upon the biosynthesis of starch.[1] The backbone model can contribute to our view of the biosynthesis of amylopectin taking into account the action of starch-branching enzymes.

REFERENCES

1. E. Bertoft, *Agronomy*, 7, 56, 2017 in http://www.mdpi.com/2073-4395/7/3/56
2. B. Pfister and S.C. Zeeman, *Cell Mol. Life Sci.*, 73, 2781, 2016 in https://link.springer.com/article/10.1007/s00018-016-2250-x
3. *Glycogen Metabolism* in http://core.ecu.edu/biol/evansc/PutnamEvans/5800pdf/GlycogenMetabolism.pdf
4. S.C.A. Allay and M.A.A. Meireles, *Food. Sci. Technol.* (Campinas) 35, 2, Campinas, 2015 in http://www.scielo.br/scielo.php?script=sci_arttext&pid=S0101-20612015000200215
5. M. Chaplin, *Water Structure and Science*, 2018 in http://www1.lsbu.ac.uk/water/starch.html
6. B.M.J. Martens, W.J.J. Gerrits, E.M.A.M. Bruininx and H.A. Schols, *J. Anim. Sci. Biotechnol.*, 9, 91, 2018 in https://www.ncbi.nlm.nih.gov/pubmed/30619606
7. A. Sarko and H.C.H. Wu, *Starch, The Crystal Structures of A-, B- and C-Polymorphs of Amylose and Starch*, Vol 30, 73, Wiley, 1978 in https://onlinelibrary.wiley.com/doi/abs/10.1002/star.19780300302
8. M.J. Gidley, *Macromolecules*, 22, 351, 1989 in https://pubs.acs.Torg/doi/abs/10.1021/ma00191a064
9. T. Luallen, <u>Starch in Food</u>, 2004 in https://www.sciencedirect.com/topics/agricultural-and-biological-sciences/amylose
10. Y. Takeda J. *Jpn. Soc. Starch Sci.*, 40, 61, 1993 in https://www.jstage.jst.go.jp/article/jag1972/40/1/40_1_61/_article
11. J.L. Wertz, O. Bedue and J.P. Mercier, *Cellulose Science and Technology*, EPFL Press, 2010
12. S.E. Dew, Chemistry Explained, Advameg, Inc., 2018 in http://www.chemistryexplained.com/Pl-Pr/Polysaccharides.html
13. S. Perez and A. Imberty, *Structure and Morphology of Starch*, 2002 in https://www.researchgate.net/publication/280819214_Structure_and_Morphology_of_Starch/link/5625469a08ae4d9e5c4bb194/download
14. S. Perez and E. Bertoft, *Starch/Stärke*, 62, 389, 2010 in https://www.researchgate.net/publication/229861904_The_molecular_structures_of_starch_components_and_their_contribution_to_the_architecture_of_starch_granules_A_comprehensive_review.
15. J.N. Bemiller and R.N. Whisler, *Starch: Chemistry and Technology*, 3rd edition, Elsevier, Academic Press, 2009 in https://www.elsevier.com/books/starch/bemiller/978-0-12-746275-2

16. J.L. Jane and J.J. Shen, *Carbohydr. Res.*, 247, 279, 1993 in https://books.google.be/boo
    ks?id=t3UoDwAAQBAJ&pg=PA7&lpg=PA7&dq=J.L.+JANE+and+J.J.+SHEN,+Car
    bohydr.+Res.+247,+279,+1993&source=bl&ots=_lgnAp4TGq&sig=ACfU3U0ZQ4gYB
    vh9p2xBaPqPGQwTpidNGQ&hl=fr&sa=X&ved=2ahUKEwi9v6-To5_oAhWN2KQK
    HZWCCEEQ6AEwAXoECAwQAQ#v=onepage&q=J.L.%20JANE%20and%20J.J.%20
    SHEN%2C%20Carbohydr.%20Res.%20247%2C%20279%2C%201993&f=false

17. D.D. Pan and J.L. Jane, *Biomacromolecules*, 1, 126, 2000 in https://books.google.
    be/books?id=abwlCQAAQBAJ&pg=PA59&lpg=PA59&dq=D.D.+PAN+and+J
    .L.+JANE,+Biomacromolecules+1,+126,+2000&source=bl&ots=oUtu45IeZE
    &sig=ACfU3U0ieIO0G5Su7xszh99-fu1lOUfeSw&hl=fr&sa=X&ved=2ahUKE
    wjs2PHOo5_oAhXNyKQKHXGsCSYQ6AEwAXoECAoQAQ#v=onepage&q=
    D.D.%20PAN%20and%20J.L.%20JANE%2C%20Biomacromolecules%201%2C%20
    126%2C%202000&f=false

18. M.A. Glaring, C.B. Koch and A. Blennow, *Biomacromolecules*, 7, 2310, 2006 in https://
    www.ncbi.nlm.nih.gov/pubmed/16903676

19. J.L. Jane, A. Xu, M. Radosavljevic and P.A. Seib, *Cereal Chem.*, 69, 405, 1992 in
    https://books.google.be/books?id=6p_fm6jNrmMC&pg=PA161&lpg=PA161&dq=J.
    L.+JANE,+A.+XU,+M.+RADOSAVLJEVIC+and+P.A.+SEIB,+Cereal+Chem.,+69,
    +405,+1992&source=bl&ots=BJb0E4xgWx&sig=ACfU3U1wgGgMmuPTI4aqMN
    PuIrI3193ZDg&hl=fr&sa=X&ved=2ahUKEwjlq-z7o5_oAhWCqaQKHeT5Ci4Q6AE
    wAHoECAIQAQ#v=onepage&q=J.L.%20JANE%2C%20A.%20XU%2C%20M.%20
    RADOSAVLJEVIC%20and%20P.A.%20SEIB%2C%20Cereal%20Chem.%2C%20
    69%2C%20405%2C%201992&f=false

20. S. Perez, *Cermav, Lessons, Starch, The* Location *and* State *of* Amylose *within the*
    Granule in http://applis.cermav.cnrs.fr/lessons/starch/page.php.22.html

21. J.N. Bemiller and R.N. Whisler, *Starch: Chemistry and Technology*, 3rd edition,
    Elsevier, Academic Press, 2009 in https://www.elsevier.com/books/starch/bemiller/
    978-0-12-746275-2

22. F. Teyssandier, These Insa de Lyon, 2011 in http://theses.insa-lyon.fr/publication/
    2011ISAL0125/these.pdf

23. C. Gernat, S. Radosta, H. Anger and G. Damaschun, *Starch/Stärke*, 45, 309, 1993 in
    https://onlinelibrary.wiley.com/doi/abs/10.1002/star.19930450905

24. *Amylose* in https://en.wikipedia.org/wiki/Amylose

25. W.C. Obiro, S.S. Ray and M.N. Emmambux, *Food Rev. Int.*, 28, 4, 2012 in https://www.
    tandfonline.com/doi/abs/10.1080/87559129.2012.660718

26. J.R. Katz and T.B. Van Itallie, Zeitschrift für Physikalische chemie, *Abteilung A*, A150,
    90, 1930 in https://books.google.be/books?id=KN7ZBkLzwRoC&pg=PA264&lpg=PA2
    64&dq=J.R.+KATZ+and+T.B.+VAN+ITALLIE,+Zeitschrift+f%C3%BCr+Physikalisc
    he+chemie,+Abteilung+A,+A150,+90,+1930&source=bl&ots=uMpqPHNebI&sig=AC
    fU3U31ScKiceKds6luQBjbN6_kdcoeEg&hl=fr&sa=X&ved=2ahUKEwj73cTDpJ_oAh
    UDjqQKHeJ2ASEQ6AEwAHoECAkQAQ#v=onepage&q=J.R.%20KATZ%20and%20
    T.B.%20VAN%20ITALLIE%2C%20Zeitschrift%20f%C3%BCr%20Physikalische%20
    chemie%2C%20Abteilung%20A%2C%20A150%2C%2090%2C%201930&f=false

27. W.O. Cuthbert, S.S. Ray, and M.N. Emmambux, Department of Food Science,
    University of Pretoria in https://repository.up.ac.za/bitstream/handle/2263/21993/
    Cuthbert_V-amylose_2012.pdf?sequence=1&isAllowed=y

28. S.G. Ring, K.J. L'anson, and V.J. Morris, *Macromolecules*, 18, 182, 1985 in https://sor.
    scitation.org/doi/abs/10.1122/1.550997

29. A.C. Eliasson, Ed, *Starch in Food, Structure, Function and Applications*, CRC Press,
    2004 in https://trove.nla.gov.au/work/8058208?q&versionId=165341 56+225898856

30. Encyclopedia.com, *Polysaccharides*, The Gale Group, Inc, 2004 in https://www.
    encyclopedia.com/science-and-technology/biochemistry/biochemistry/
    polysaccharides

31. M. Sjoo, L. Nilsson, Eds, *Starch in Food: Structure, Function and Applications*, 2nd edition, Elsevier, Woodhead Publishing, 2018 in https://books.google.be/books?id=gyx HDgAAQBAJ&hl=fr&redir_esc=y

32. E. Bertoft, The unit chain profile of amylopectins – an internal view, *Starch Convention*, Detmold, 2008 in https://onlinelibrary.wiley.com/doi/abs/10.1002/star.200890053

33. D.N. Kalinga, A thesis presented to the University of Guelph, Canada, 2013 in https://www.lib.uoguelph.ca/find/find-type-resource/theses-dissertations

34. S. Peat, W.J. Whelan, G.J. Thomas, *J. Chem. Soc.*, 4546, 1952 in https://miami.pure.elsevier.com/en/publications/evidence-of-multiple-branching-in-waxy-maize-starch

35. Y. Nakamura, Ed., *Starch, Metabolism and Structure*, Springer, 2015 in https://www.springer.com/us/book/9784431554943 and https://link.springer.com/book/10.1007%2F978-4-431-55495-0

36. A. Bijttebier, H. Goesaert and J. A. Delcour, *Biologia*, 63, 989, 2008 in https://link.springer.com/article/10.2478/s11756-008-0169-x

37. D. Perin and E. Murano, *Nat. Commun.*, 12, 837, 2017 in https://www.researchgate.net/figure/Structures-of-amylopectin-Dots-represent-glycosyl-residues-horizontal-lines-represent_fig1_317949581 and in https://www.researchgate.net/publication/317949581_Starch_Polysaccharides_in_the_Human_Diet_Effect_of_the_Different_Source_and_Processing_on_its_Absorption?_sg=f1Z0UQM_FXbapAvuOLyi-v-6Q0NPAhCOyf9TI8Tti-X4CtfViLIoFLDr4BfIjzgJz1xVmxJLQA

38. V. Vamadevan and Q. Liu, *Carbohydrates, Starch: Starch Architecture and Structure*, p. 193, 2016 in https://www.researchgate.net/publication/292783871_Starch_Starch_Architecture_and_Structure

39. E. Bertoft, A. Kallman, K. Koch, R. Andersson and P. Aman, Int. J. Biolog. *Macromolecules*, 49, 900, 2011 in https://www.sciencedirect.com/science/article/pii/S0141813011003011

40. E. Bertoft, K. Koch and P. Aman, *Carbohydr. Res.*, 361, 105, 2012 in https://www.sciencedirect.com/science/article/abs/pii/S000862151200359X

41. Z. Nikuni, *Sci Cookery*, 2, 6, 1969 in https://books.google.be/books?id=Rp LMBQAAQBAJ&pg=PA57&lpg=PA57&dq=Z.+NIKUNI,+Sci+Cookery,+2,+6,+1969&source=bl&ots=tEn_oa4Ogl&sig=ACfU3U2-sbutvY8wD-FAp9InwcO-8dBRZHw&hl=fr&sa=X&ved=2ahUKEwjN-P7qpp_oAhWP-aQKHf2LBOwQ6AEwCnoECAsQAQ#v=onepage&q=Z.%20NIKUNI%2C%20Sci%20Cookery%2C%202%2C%206%2C%201969&f=false

42. D. French. *J. Jpn. Soc. Starch Sci.*, 19, 8, 1972 in https://www.sciencedirect.com/science/article/abs/pii/S002253207990114X

43. S. Hizukuri, *Carbohydr. Res.*, 147, 342, 1986 in https://www.sciencedirect.com/science/article/abs/pii/S0008621500906438

44. L. Hanashiro, J.L. Abe, S. Hizukuri, *Carbohydr. Res.*, 283, 151, 1996 in https://books.google.be/books?id=JCTNDwAAQBAJ&pg=PA206&lpg=PA206&dq=L.+HANASHIRO,+J.L.+ABE,+S.+HIZUKURI,+Carbohydr.+Res.+283,+151,+1996&source=bl&ots=I7dbmaRmSV&sig=ACfU3U2JEOH6DSARMAHHZ_B-5lXFbwGDSA&hl=fr&sa=X&ved=2ahUKEwjMpLrit5_oAhWS6qQKHX19AK4Q6AEwC3oECAoQAQ#v=onepage&q=L.%20HANASHIRO%2C%20J.L.%20ABE%2C%20S.%20HIZUKURI%2C%20Carbohydr.%20Res.%20283%2C%20151%2C%201996%2C%201996&f=false

45. A.C. O'sullivan and S. Perez, *Biopolymers*, 50, 381, 1999 in https://www.ncbi.nlm.nih.gov/pubmed/10423547

46. K. Laohaphatanaleart, K. Piyachomkwan, K. Sriroth, and E. Bertoft, *Int. J. Biol. Macromol.*, 47, 317, 2010 in https://www.ncbi.nlm.nih.gov/pubmed/20083134

47. E. Bertoft, *Cereal Chem.*, 90, 294, 2013 in https://books.google.be/books?id-=abwlCQAAQBAJ&pg=PA34&lpg=PA34&dq=E.+BERTOFT,+Cereal+Chem.+90,+294,+2013&source=bl&ots=oUtu457a1F&sig=ACfU3U2FtYU4Pr

PAeYniIDT6KallIGiEnA&hl=fr&sa=X&ved=2ahUKEwj7z4jRuJ_oAhWD-6QKHWuQC5EQ6AEwAXoECAwQAQ#v=onepage&q=E.%20BERTOFT%2C%20Cereal%20Chem.%2090%2C%202094%2C%202013&f=false

48. E. Bertoft, *Carbohydr. Polym.*, 57, 211, 2004
49. A. Kallman, E. Bertoft, K. Koch, P. Aman and R. Andersson, *Int. J. Biol. Macromol.*, 55, 75, 2013 in https://www.ncbi.nlm.nih.gov/pubmed/23270830
50. E. Bertoft, K. Koch and P. Aman, *Int. J. Biol. Macromol.*, 50, 1212, 2012 in https://www.ncbi.nlm.nih.gov/pubmed/22465108
51. A. Kallman, E. Bertoft, K. Koch, P. Aman and R. Andersson, *Macromolecules*, 55, 75, 2013 in https://www.ncbi.nlm.nih.gov/pubmed/23270830
52. S. Perez and E. Bertoft, *Starch-Starke*, 62, 389, 2010 in https://www.researchgate.net/publication/229861904_The_molecular_structures_of_starch_components_and_their_contribution_to_the_architecture_of_starch_granules_A_comprehensive_review
53. F. Spinozzi, C. Ferrero and S. Perez, *Sci. Rep.*, 2020 (in press)
54. L. Taiz, E. Zeiger, I.M. Moller and A. Murphy, *Plant Physiology and Development*, 6th edition, Topic 8.13 Starch Architecture, Sinauer Associates, 2015 in http://6e.plantphys.net/topic08.13.html
55. S. Perez and A. Imberty, *Structure and Morphology of Starch*, 2002 in https://www.researchgate.net/publication/280819214_Structure_and_Morphology_of_Starch
56. E. Bertoft, Starch in Food, 2nd edition, *Structure, Function and Applications*, 97, 2018 in https://www.sciencedirect.com/book/9780081008683/starch-in-food
57. A. Buleon, D.J. Gallant, B. Bouchet, G. Mouille, C. D'hulst, J.Kossmann and S. Ball, *Plant Physiol.*, 115, 949, 1997 in http://www.plantphysiol.org/content/plant-physiol/115/3/949.full.pdf
58. J. Ahmed, B.K. Tiwari, S.H. Imam and M.A. Rao Eds., *Starch-Based Polymeric Materials and Nanocomposites*, CRC Press, 2012 in https://www.crcpress.com/Starch-Based-Polymeric-Materials-and-Nanocomposites-Chemistry-Processing/Ahmed-Tiwari-Imam-Rao/p/book/9781138198623
59. D Popov, M. Burghammer, A. Buleon, N. Montesanti, J.L. Putaux, and C. Riekel, *Macromolecules*, 39, 3704, 2006 in https://pubs.acs.org/doi/abs/10.1021/ma060114g
60. D. Popov, A. Buleon A, M. Burghammer, H. Chanzy, N. Montesanti, J.L. Putaux, G. Potocki-Veronese and C. Riekel. *Macromolecules*, 42, 1167, 2009 in https://pubs.acs.org/doi/10.1021/ma801789j
61. A. Imberty and S. Perez, *Biopolymers*, 27, 1205, 1988 in https://onlinelibrary.wiley.com/doi/pdf/10.1002/bip.360270803
62. Y. Takahashi, T. Kumano and S. Nishikawa, *Macromolecules*, 37, 6827, 2004 in https://pubs.acs.org/doi/10.1021/ma0490956
63. T. A. Waigh, I. Hopkinson, A. M. Donald, M. F. Butler, F. Heidelbach and C. Riekel, *Macromolecules*, 30, 3813, 1997 in https://scholar.google.co.uk/citations?user=FY0pJcYAAAAJ&hl=en#d=gs_md_cita-d&u=%2Fcitations%3Fview_op%3Dview_citation%26hl%3Den%26user%3DFY0pJcYAAAAJ%26citation_for_view%3DFY0pJcYAAAAJ%3AkNdYIx-mwKoC%26tzom%3D-120
64. A. Buleon, C. Gerard, C. Riekel, R. Vuong and H. Chanzy, *Macromolecules*, 31, 6605, 1998 in https://pubs.acs.org/doi/abs/10.1021/ma970136q
65. A. Dufresne, S. Thomas and L.A. Pothan Eds., *Biopolymers Nanocomposites, Processing, Properties and Applications*, Wiley, 2013.
66. M. Perez-Herrera, T. Vasanthan and R. Hoover, *Cereal Chem.*, 93 (3), 2015 in https://www.researchgate.net/publication/286401283_Characterization_of_Maize_Starch_Nanoparticles_Prepared_by_Acid_Hydrolysis
67. K.W. Von Nageli in https://fr.wikipedia.org/wiki/Karl_Wilhelm_von_N%C3%A4geli

3 Biosynthesis of Starch

3.1 INTRODUCTION

3.1.1 ADPglucose and Fundamentals of Starch Synthesis

The *ADPglucose* (adenosine diphosphate glucose, Figure 3.1) pathway, comprising the enzymatic reactions catalyzed respectively by ADPglucose pyrophosphorylase (AGPase; EC 2.7.7.27), (Equation 3.1), starch synthase (SS) (Equation 3.2), and branching enzyme (BE) (Equation 3.3), is the predominant pathway toward starch synthesis in leaf as well as in non-photosynthetic storage tissues such as endosperm or tubers.[1,2,3,4]

α-Glucosyl-1-P (abbreviated to G1P) + ATP $\rightleftarrows$ ADPglucose + PP$_i$ (catalyzed by AGPase) (Equation 3.1)

ADPglucose + (glucosyl)$_n$ → ADP + (glucosyl)$_{n+1}$ (catalyzed by SS) (Equation 3.2)

Linear glucan → Branched glucan with α-(1,6)-linkage (catalyzed by BE) (Equation 3.3)

ADP is a *nucleotide*, consisting of a nucleoside (composed of a nitrogenous base and a five-carbon sugar) and one to three phosphate groups. ADPglucose is an activated form of glucose.[5] It is utilized as the glucosyl donor for the elongation of the α-1,4-glucosidic chain.[6] α-Glucosyl-1-P used in Equation (3.1) with *ATP* to form

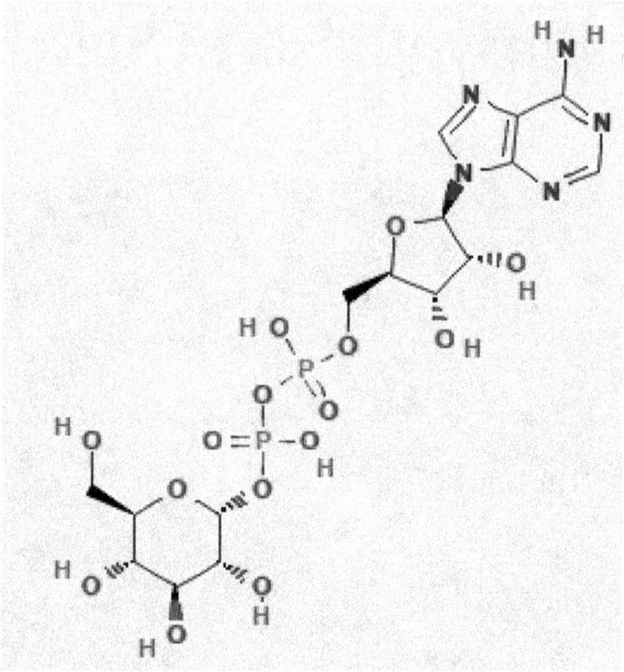

FIGURE 3.1   ADPglucose (adenosine diphosphate glucose). (Courtesy of the U.S. National Library of Medicine.)[7]

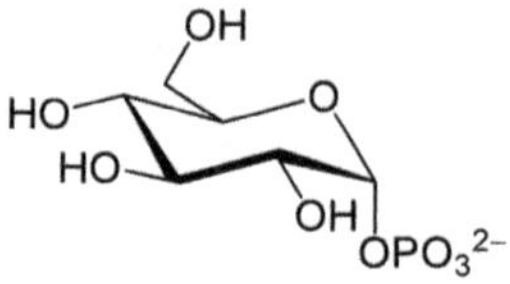

FIGURE 3.2   α-Glucosyl-1-P also called glucose-1-phosphate (G1P).

ADPglucose is commonly called glucose-1-phosphate or G1P (Figure 3.2). It contains the phosphate group in position 1 of the glucose ring.

From our current understanding of starch biosynthesis from ADPglucose, the process includes (1) granule-bound starch synthase (GBSS)[8] synthesizing amylose and (2) soluble SSs, BEs, and isoamylase-type debranching enzymes (ISA) collectively synthesizing amylopectin (Figure 3.3).[9]

3.1.2   Starch Synthesis from the Calvin Cycle

Starch is stored in *plastids* of essentially all plant tissues and consumed as both energy and carbon source.[3] During photosynthesis, plants produce storage compounds mainly in the form of starch and *sucrose* from carbon dioxide and water. Carbon dioxide is first incorporated into the intermediates in the Calvin cycle, also referred to as the Calvin–Benson cycle.

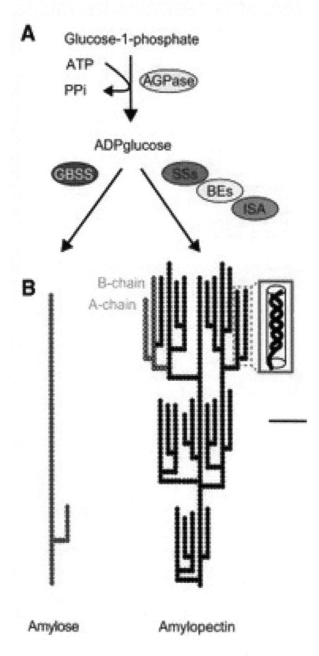

FIGURE 3.3 Biosynthesis of starch. (a) Overview of the core starch biosynthesis pathway. ADPglucose pyrophosphorylase (AGPase) produces ADPglucose, the substrate of starch synthases (SSs). Granule-bound starch synthase (GBSS) synthesizes amylose, whereas soluble SSs, branching enzymes (BEs), and isoamylase-type debranching enzyme (ISA) collectively synthesize amylopectin. (b) Molecular structure of amylose and amylopectin (according to the cluster model), showing its branching pattern. (Reproduced with permission from Springer Nature.)[9]

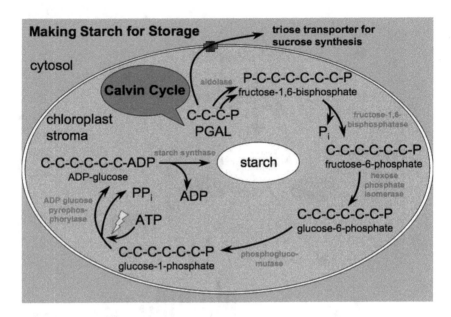

FIGURE 3.4 Formation of starch in photosynthetic tissues from the Calvin cycle. (Reproduced with the permission of R.E. Koning.)[14]

The Calvin cycle[10] includes three stages: (1) carbon fixation, where a carbon dioxide molecule combines with a five-carbon acceptor molecule to form a six-carbon molecule that splits into two molecules of 3-phosphoglycerate (3-PGA); (2) reduction, where the 3-PGA molecules are converted into molecules of glyceraldehyde-3-phosphate, also known as triose phosphate or 3-phosphoglyceraldehyde and abbreviated as G3P or PGAL; and (3) regeneration, where some G3P molecules go to make glucose, while others are recycled to generate the initial five-carbon acceptor. In photosynthetic tissues, starch and sucrose are synthesized from G3P in the *chloroplast* and *cytosol* respectively (Figure 3.4).[4,11–14]

3.1.3 TRANSITORY AND STORAGE STARCH

Starch is the predominant reserve form of carbohydrate and energy in plants and can be divided into two types, transitory starch and storage starch, based on biological function.[15] In photosynthetic tissues such as leaves (also referred to as source organs), transient starches accumulate in chloroplasts during the day. During the night, they are then transported and degraded to provide energy and nutritional substances for growth and metabolism. In non-photosynthetic tissues, such as seed endosperm, tubers, and storage roots (also referred to as sink organs), storage starches are kept for long periods in specialized plastids termed *amyloplasts*, from which they can be remobilized in the preparation for germination, sprouting, or regrowth.[16] Sucrose synthesized in leaves is imported by storage tissues for starch synthesis in amyloplasts (Figure 3.5).

Starch synthesis in plants involves four major enzyme activities: first, AGPase produces ADPglucose; second, SSs elongate the non-reducing ends of glucose chains using ADPglucose as the glucosyl donor; third, BEs create branches from existing chains

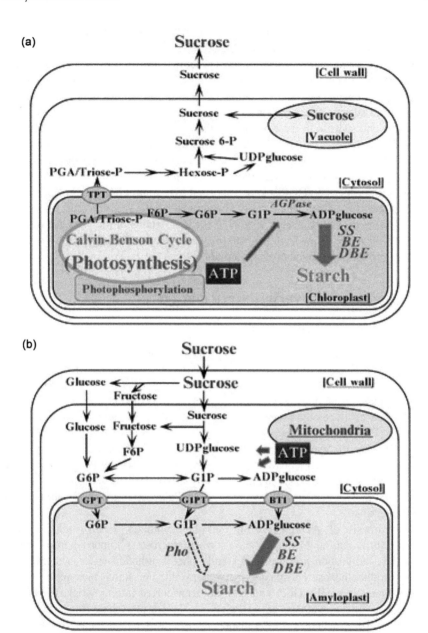

FIGURE 3.5 Schematic representation of the starch biosynthetic pathway and the related metabolism in **photosynthetic (a)** and **non-photosynthetic tissues (b)**. **(a)** The synthesis of transitory starch in leaves: starch is synthesized in chloroplasts through the Calvin cycle. **(b)** The synthesis of reserve starch in storage tissues: starch is synthesized in amyloplasts from sucrose. BT1, brittle-1 protein (ADPglucose transporter); TPT, triose phosphate/P$_i$ translocator; GPT, G6P/P$_i$ translocator; G1PT, putative G1P transporter; PGA, phosphoglycerate. (Reproduced with permission from the American Society of Plant Physiologists.)[2,3]

via glucanotransferase reactions; and fourth, debranching enzymes (DBEs) hydrolyze some of the branches again.[16] Although presented here sequentially, it is important to perceive it as a simultaneous, interdependent process. The starch-biosynthetic enzymes are well conserved between different plant species, suggesting a common origin.

During photosynthesis, plants accumulate storage compounds mainly in the form of starch and sucrose from carbon dioxide and water. Starch is synthesized in the chloroplast and sucrose in the cytosol (see Figure 3.4). ATP and NADH needed for the syntheses of these compounds are provided from the photophosphorylation system and the light-dependent electron-transport system of photosynthesis, respectively. Thus, the ratios of inorganic phosphate (P_i)/total P, ATP/ADP, and NADPH/NADP$^+$ in chloroplast and cytosol are important to maintain high rates of photosynthesis and *homeostasis* of whole cells. The photosynthetic reaction must occur very quickly to support all the energy-requiring events in plants in the presence of light. Therefore, the carbon flow from the Calvin cycle to the starch and sucrose syntheses forms the primary metabolic process in plant cells.

3.2   ADPGLUCOSE PYROPHOSPHORYLASE

3.2.1   SYNTHESIS OF ADPGLUCOSE

In photosynthetically active chloroplasts of leaves (Figure 3.5a), the generation of ADPglucose is directly linked to the Calvin–Benson Cycle through the conversion of fructose-6-phosphate to glucose-6-phosphate (G6P) (catalyzed by phosphoglucose isomerase) through to G1P (catalyzed by phosphoglucomutase).[14,17] AGPase (EC 2.7.7.27) then catalyzes the conversion of G1P and ATP to ADPglucose and pyrophosphate (PP_i). Through this pathway, approximately 30%–50 % of *photoassimilates* of *Arabidopsis* leaves are partitioned into starch. Each of the aforementioned reactions is thermodynamically reversible. However, in vivo, the PP_i product of the last reaction is further metabolized; it is hydrolyzed to yield two molecules of orthophosphate (P_i). This renders the synthesis of ADPglucose in the chloroplast essentially irreversible.

The synthesis of ADPglucose in non-photosynthetic, *heterotrophic* tissues (Figure 3.5b) is similar to that in leaves, where sucrose is imported from *source tissues* (i.e., net carbon producers, generally photosynthetic tissues) and metabolized to produce hexose phosphates in the cytosol.[13] For starch biosynthesis, hexose phosphates (typically G6P) and ATP are transported into amyloplasts to serve as substrates for the synthesis of ADPglucose. Hexose phosphate transport occurs in exchange for P_i, whereas ATP transport occurs in exchange for ADP and P_i. In the cereal endosperm, the pathway differs: the major AGPase activity is found in the cytosol and ADPglucose is imported directly into the plastid via a dedicated transporter.

The synthesis of starch and sucrose plays an important role in photosynthetic CO_2-fixation activity in returning P_i to the photophosphorylation system and the Calvin cycle. Because sucrose is synthesized in the cytosol, the released P_i is returned to chloroplasts by the triose-P/P_i translocator located in the inner envelope of chloroplasts when carrying triose-P/PGA into the cytoplasm (see Figure 3.5a).

3.2.2 CRYSTAL STRUCTURE OF AGPASE

AGPase catalyzes the first committed and rate-limiting step in starch biosynthesis in plants and glycogen biosynthesis in bacteria.[18] There is great evidence that the step is regulated both at the *transcriptional* and post-*translational* levels.[9] Although the overall kinetic mechanism of AGPase appears to be similar in bacteria and higher plants, their quaternary structures differ from each other.[19] Bacterial AGPases are composed of four identical subunits (α) to form $\alpha 4$ homotetramer, whereas plant AGPases are heterotetramer of two different yet evolutionarily related subunits containing a pair of identical small subunits (α) and identical large subunits (β) to form $\alpha 2 \beta 2$ heterotetramer. The two subunits vary in their molecular weight and genetic origin and are encoded by two different genes. Primary sequence analysis of large and small subunits of AGPase has shown considerable sequence homology, suggesting a common evolutionary origin. The small subunit of AGPase has both catalytic and regulatory functions, whereas the large subunit has only a regulatory function.[20,21] The small subunit is capable of forming a homotetramer with catalytic properties, whereas the large subunit is incompetent of forming an oligomeric structure with catalytic activities. In contrast, some researchers suggested that the large subunit might interact with the catalytic small subunit influencing the net catalysis. Consequently, both the subunits are of equal importance for the catalysis and allosteric regulation of the enzyme.

In 2005, Jin et al.[18] reported the first atomic-resolution structure of the small subunit of AGPase from potato tuber (PDB ID: 1YP4, Figure 3.6). The crystal structure of the small subunit was found in a homotetrameric form.

3.2.2.1 Monomer

The α-subunit monomer of potato tuber AGPase is composed of an N-terminal catalytic domain and a C-terminal β-helix domain (Figure 3.6a).[27] The overall fold of the catalytic domain shares a strong similarity with the three PPis whose structures have been elucidated, namely N-acetyl glucosamine-1-phosphate uridyltransferase (GlmU) and two G1P thymidylyltransferases (RmlA and RffH), although the primary sequences for the three distinct enzymes have very low similarities (Figure 3.6b). The catalytic domain is composed of a mostly parallel but mixed seven-stranded β-sheet covered by α-helices, a fold reminiscent of the dinucleotide-binding *Rossmann fold*. The catalytic domain makes strong hydrophobic interactions with the C-terminal β-helix domain via an α-helix that encompasses residues 285–297. The catalytic domain is connected to the C-terminal β-helix domain by a long loop containing residues 300–320. This loop makes numerous interactions with the equivalent region of another subunit.

The C-terminal domain comprising residues 321–451 adopts a left-handed β-helix fold composed of six complete or partial coils with two insertions, one of which encompasses residues 368–390 and the other encompasses residues 401–431. This type of left-handed β-helix domain fold has been found in the structures of several proteins but the orientation and length of the β-helix domain are completely different (Figure 3.6b). Functionally, the other β-helix domains are either acetyltransferases or succinyltransferases. However, in AGPase, the β-helix domain is involved in

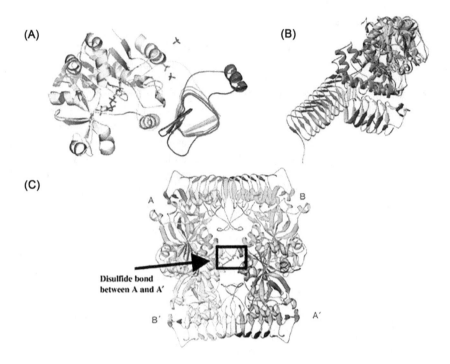

FIGURE 3.6 (a) AGPase α-subunit monomer. Yellow, catalytic domain; pink, β-helix domain; ADP-glucose and sulfates are shown in atom type: carbon, green; oxygen, red; nitrogen, blue; phosphorous, magenta; sulfate, orange. (b) Overlay of AGPase (cyan), *N*-acetylglucosamine 1-phosphate uridyltransferase RmlA (magenta), and G1P thymidylyltransferases GlmU (gold). (c) AGPase α₄ tetramer. The disulfide bond between A and A' is boxed. (Reproduced with full access to the article.)[18]

cooperative allosteric regulation with the N-terminal catalytic region, interacts with the N-terminus within each monomer, and contributes to the unique oligomerization unprecedented in other structures.

3.2.2.2 Tetramer

The potato tuber AGPase α₄-homotetramer has approximate 222 symmetry[22,23] with dimensions of ~80×90×110 Å³; it can be viewed as a dimer of dimers with each monomer labeled as A, A', B, and B' (Figure 3.6c).[26] Monomers A and B interact predominantly by an end-to-end stacking of their β-helix domains, although there is also a significant interface between the linker loop connecting the N-terminal domain and C-terminal domain (residues 300–320). This interface buries a surface area of 2,544 Å². The catalytic domains of the A and B' (and B and A') subunits also make an extensive interface. Several hydrogen bonds and hydrophobic interactions stabilize this interface by burying a surface area of 1,400 Å². All of the residues defining the dimerization interfaces are identical or similar in the β-subunit, suggesting a similar dimerization interface between α- and β-subunits in the heterotetrameric enzyme (Figure 3.6c). Cys12 of monomer A and the equivalent cysteine residue of monomer

A' make a disulfide bond, as do equivalent cysteines of monomers B and B', and this is the only interaction between A and A' (B and B'). The intersubunit disulfide bond between α-subunits is preserved in the heterotetramer; however, there is no disulfide bond between β-subunits as Cys12 is not conserved. This disulfide bond establishes the relative position of α-subunits in the heterotetramer to be like A and A' in the α_4 homotetramer structure.

3.2.2.3   Activation and Inhibition

In many cases, AGPase has been demonstrated to be *allosterically* activated by 3-PGA and inhibited by P_i.[6] The enzyme is furthermore sensitive to redox regulation via the reduction of an intermolecular disulfide bridge that forms between cysteine residues of the small subunit. Together, these regulatory features are thought to ensure that ADPglucose, and thus starch, is only synthesized when there are sufficient substrates. Many attempts have been made to promote the flux toward starch by expressing unregulated AGPase from *Escherichia coli* or plants. This has resulted in an increased starch content in at least one potato variety, increased overall grain yield in maize and wheat, and increased tuberous root biomass in cassava.

3.3   STARCH SYNTHASES

SSs (EC 2.4.1.21) belong to the retaining *glycosyltransferase (GT)* family 5 (CAZy).[6, 24] GT5 also includes the ADPglucose-utilizing bacterial glycogen synthases (GSs). SSs catalyze the transfer of the glucosyl moiety of ADPglucose to the non-reducing end (the C4 position) of an existing glucosyl chain, creating an α-1,4 bond and elongating the chain. Five SS classes are involved in starch biosynthesis: four are soluble in the *stroma* of the chloroplast (or partially bound to the granule) and one is almost exclusively granule bound. The soluble SSs (SSI, SSII, SSIII, and SSIV) are involved in amylopectin synthesis while the GBSS is responsible for amylose synthesis. There is an additional putative SS class named SSV that is related in sequence to SSIV.

SSs consist of a highly conserved C-terminal catalytic domain and a variable N-terminal extension (Figure 3.7).[9] The catalytic domain is conserved between SSs and bacterial GSs and contains both a GT5 domain and a GT1 domain (CAZy).[33] According to the crystal structures of GSs of *Agrobacterium tumefaciens* and *E. coli*, the rice GBSSI, and barley SSI, the catalytic domain adopts a *GT-B fold*, with the active site in a cleft between the two GT domains. Binding of ADPglucose probably involves one or more conserved Lys-X-Gly-Gly (lysyl-residue not conserved-glycyl-glycine) sequences and other conserved charged/polar residues.[25] The N-terminal extensions of SS classes are dissimilar. In the case of SSIII and SSIV, these extensions were shown to be involved in protein–protein interactions, potentially via conserved *coiled-coil* motifs. The N-terminal part of SSIII also contains three conserved carbohydrate-binding modules (CBMs) that are involved in substrate binding.

Gene duplications have resulted in multiple isoforms of some enzymes.[15] The encoded proteins have a high degree of sequence similarity but are often differentially expressed, with specific isoforms predominating in the endosperm or vegetative tissues. The "a" isoforms of SSII and SSIII appear to be the predominant

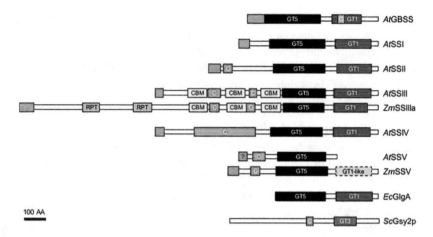

FIGURE 3.7 The domain structure of starch synthase (SS) classes. SSs from *Arabidopsis* (*At*) compared with glycogen synthases from *E. coli* (EcGlgA) and budding yeast (ScGsy2p). Maize (*Zm*) SSIIIa and SSV are included as they differ in their structures compared with the *Arabidopsis orthologs*. Shown are plastidial *transit peptides* (N-terminal blue bases), internal repeats (gray boxes, RPT domain), carbohydrate-binding modules of family 25 (yellow boxes, CBM), coiled-coil domains (green boxes, C), glycosyltransferase-5 domains (black boxes, GT5), glycosyltransferases-1 domains (red boxes, GT1), and a glycosyltranferase-3 domain (orange box, GT3). Bar 100 amino acids (AA). (Reproduced with permission from Springer Nature.)[9]

isoforms in the endosperm. In other species, including *Arabidopsis* and plants having storage starch-filled organs (e.g., potato), there is only one isoform for each class.

Recently, Nielsen et al.[26] reported that the crystal structures of the catalytic domain of SSIV from *Arabidopsis thaliana*, GBSS from the cyanobacterium CLg1, and GBSSI from the unicellular glaucophyte *Cyanophora paradoxa* illustrate substrate recognition in SSs.

3.3.1 SYNTHESIS OF AMYLOSE BY GRANULE-BOUND STARCH SYNTHASE

3.3.1.1 Introduction

Luis Leloir's group initially discovered SS activity that was associated with the starch granule.[19] The original characterization of GBSS was made using UDPglucose (uridine diphosphate glucose) as the glucosyl donor. Subsequently, ADPglucose was found to be a superior glucosyl donor.

Mutants with reduced or no GBSS activity, the so-called *waxy* lines, produce less or no amylose for instance in the endosperms of maize, rice, wheat, barley, amaranth, cassava roots, potato, pea seeds, *Arabidopsis* leaves, and *Chlamydomonas reinhardtii* (single-cell green alga).[15] This suggests that GBSS is responsible for the synthesis of amylose and that no other synthase can replace it in this function.

It is likely that GBSS synthesizes amylose within the granular matrix formed by amylopectin.[6] Monitoring the distribution of amylose over time in potato lines, in which GBSS expression and amylose contents were repressed to low levels, suggested that amylose was more apparent toward the center of the starch granule and that this amylose-containing core grows together with the granule. It is important to realize that the granule is hydrated and small molecules such as ADPglucose can be used by granule-bound proteins. GBSS acts in a processive manner. The enzyme may synthesize amylose by elongating individual glucan chains in the environment surrounding crystalline or crystallizing amylopectin. Its product probably is well protected from the branching activity, explaining why it is largely linear. The nature of the primer used for amylose synthesis is not fully resolved.

GBSS does not contain any predicted starch-binding domains (see Figure 3.7).[15] Recently, a conserved starch-binding protein equipped with a family 48 CBM and a long coiled-coil motif was shown to interact with *Arabidopsis* GBSS via a short coiled-coil motif on GBSS. This interaction was required both for efficient granule binding of GBSS and for amylose synthesis.

It appears that GBSS also contributes to amylopectin synthesis.[6] In some studies of waxy mutants, the structure of amylopectin was reported to be slightly altered, whereas in other studies, it appeared normal. A notable exception is *C. reinhardtii*, in which the lack of GBSS caused an alteration in the amylopectin structure. The distinctive function of *C. reinhardtii* GBSS compared with GBSSs from vascular plants may be explained by the presence of a unique C-terminal tail and the several-fold higher specific activity.

3.3.1.2 Crystal Structure of CLg1GBSS

The crystal structure of GBSS from the cyanobacterium CLg1 contains three crystallographically independent protein molecules in the asymmetric unit, of which one (chain A) is bound to ADP (donor) and *acarbose* (acceptor), one (chain B) is bound to ADP and glucose from the cryoprotectant, and the third chain appears to be a mixture of the two.[26] The overall structure of chains A and B is shown in Figure 3.8. The structure reveals the characteristic *GT-B fold* with distinct N-terminal and C-terminal Rossmann fold domains, an interdomain linker, and a crossover helix at the C-terminus linking to the N-terminal subdomain. Two loops were disordered and not modeled in chain B, and its glucose is bound in a position equivalent to that of the last hexose, the amino-pyranose, of acarbose in chain A. Acarbose is bound with its non-reducing end hexose (4-amino-4,6-dideoxy-D-glucopyranose, also called amino-pyranose) in close contact with the phosphate moieties of ADP. Amino-pyranose mimics the glucose transferred from the donor in the reaction, while the other three acarbose hexoses occupy the binding sites of the acceptor glucan chain (chain A). The structure is representative of the conformation of the productive ternary complex among enzyme, sugar donor, and acceptor even though ADP is a reaction product.

Chain A and chain B could be described as structures in the presence and absence of an acceptor. Changes in the conformation of the protein accompany acceptor binding (Figure 3.9) and presumably precede catalytic activity.

3.3.2 THE CORE AMYLOPECTIN SYNTHESIS: STARCH SYNTHASES SSI TO SSIII

Based on mutant phenotypes, each SS class appears to have a distinct role during amylopectin synthesis; this is described simply as follows: SSI and SSII are thought

(A)

(B)

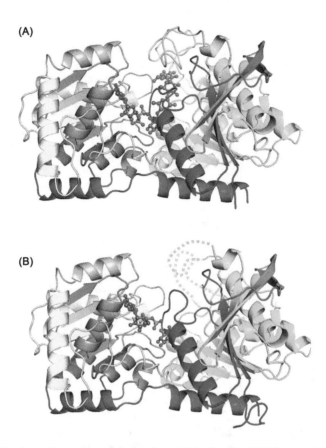

FIGURE 3.8 Overall structure of chains A and B in the CLg1GBSS crystal. The proteins are represented as ribbons with colors from dark blue at the N-terminus through to red at the C-terminus. Ligands (except water and sulfates) are shown in ball and stick with gray carbons. (a) Structure of chain A in the crystal, with ADP and acarbose (acceptor) bound and no missing loops. (b) Structure of chain B in the crystal, with ADP and glucose bound; loops missing in the model are depicted as dashed lines. (Open-access article distributed under the terms of the Creative Commons Attribution License (CC BY).)[26]

to produce the short single-cluster-filling chains (i.e., the A- and B_1-chains) while SSIII is proposed to synthesize longer cluster-spanning B chains.[15] In contrast, SSIV appears to be less involved in the determination of the amylopectin structure but to function in starch granule initiation and the control of granule morphology. However, the situation is more complicated: there are instances of functional overlap between enzymes, interferences in biosynthesis by starch degradative enzymes, and complex formation between enzymes. Besides, how each class fulfills its proposed role at the molecular level is not completely understood.

The relative contribution of each SS class varies in different tissues and between species, which is believed to explain the structural variation between starches from different sources. In the maize endosperm, SSI and SSIII constitute the major apparent soluble SS activities, whereas in maize leaves, no SSI transcript was detected. In contrast, SSII and SSIII are the major apparent soluble SS in the pea seed and

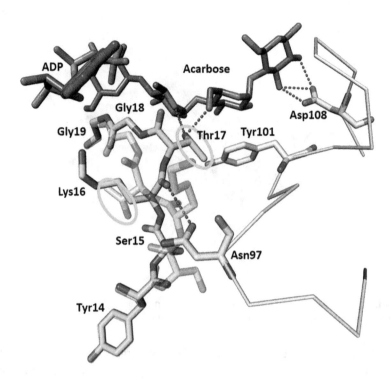

FIGURE 3.9 Changes in the conformation of CLg1GBSS accompanying acceptor binding. ADP and acarbose (the acceptor) from chain A of the CLg1GBSS crystal are shown as sticks with gray carbons. The conformation of chain A from Tyr14 to Gly19 is shown as a stick model with green carbons, while the conformation of the same segment in chain B is shown semi-transparent and with pink carbons. The loop around Tyr101 is shown as a cyan thin ribbon for the α-carbon trace with key amino acids as sticks with cyan carbons. The whole loop is disordered and absent from the model for chain B. Purple dashes represent interactions at less than 3.2 Å. Tyr101 is also participating in a stacking interaction with a glucosyl moiety in acarbose. The orange circles highlight the movement of the side chain of Thr17 upon acarbose binding. (Open-access article distributed under the terms of the Creative Commons Attribution License (CC BY).)[26]

potato tuber, whereas transcripts of potato SSI were almost exclusively detected in leaves. In *Arabidopsis* leaves, SSI is the major soluble SS, followed by SSIII and SSII. Although expressed to a reasonable level, SSIV appears to contribute only little to total SS activity.

3.3.2.1   Starch Synthase I (SSI)

Loss of SSI activity causes distinct alterations in the chain length distribution of amylopectin, particularly concerning the A- and B$_1$-chains that make up the clusters.[6] It is likely that SSI elongates the short glucan chains derived from the action of BEs (which generally have a degree of polymerization, DP, of 6) by a few glucan units (to a DP of around 8–10). These chains are then probably further elongated by SSII and possibly other SSs. However, since the majority of chains from BE action are still elongated in *ssI* mutants, the other SSs appear to be only partly dependent on the action of SSI.

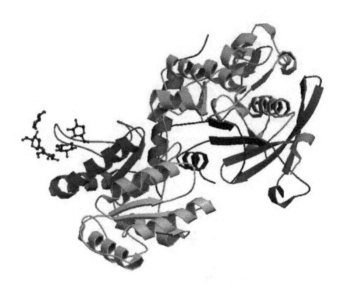

FIGURE 3.10 Structure of barley starch synthase I in complex with maltooligosaccharide; RCSB PDB 4HLN. (Reproduced with permission from RCSB.)[28]

The first crystal structure reported of a soluble SS is that of starch synthase I (SSI) from barley refined to 2.7 Å resolution (Figure 3.10).[27] The structure shows an open conformation of the enzyme with a surface-bound maltooligosaccharide and a disulfide bridge that precludes the formation of the active site. The maltooligosaccharide-binding site is involved in substrate recognition, while the disulfide bridge is reflective of redox regulation of SSI.

3.3.2.2 Starch Synthase II (SSII)

The effects of SSII deficiency have been characterized in potato tubers, pea seeds, the endosperms of wheat, barley, rice, and maize, and *Arabidopsis* leaves.[6] The observed phenotypes are remarkably similar and include a distinct change in the amylopectin fine structure: there is an increased abundance of chains around DP 8 and decreased abundance of those around DP 18, i.e., a shift toward shorter chain lengths. In addition, *ssII* mutant starches often have more amylose, altered granule morphology, and reduced starch crystallinity. In *Arabidopsis*, small amounts of soluble glucan were also reported to accumulate alongside the starch.

Based on the alterations in chain length distribution, it appears that SSII elongates chains of around DP 8 (the chains elongated by SSI) to lengths around DP 18. This direct interpretation is complicated by the fact that, at least in cereals, SSII interacts with SSI and class II BEs. Hence, the loss of SSII may have pleiotropic effects (i.e., producing more than one effect) on these enzymes, making it difficult to assess how much of the phenotype is shown directly due to the absence of SSII activity. For instance, the modest increase in amylose in *ssII* mutants might be caused by altered BEII activity. However, the changes in amylopectin fine structure are likely caused by the lack of SSII activity.

3.3.2.3 Starch Synthase III (SSIII)

Compared with SSI and SSII, the function of SSIII is less clear.[6] Its suggested functions include the synthesis of long B chains, the elongation of cluster-filling chains (partly redundant function with SSII), and the regulation of other starch-biosynthetic enzymes. Furthermore, SSIII is important for the initiation of starch granules, at least in the absence of SSIV. Consistent with the view of a versatile role, SSIII appears as a major soluble SS activity in all plants and tissues that have been analyzed to date. It also harbors the longest N-terminal extension among all SS, which carries starch-binding domains and predicted coiled-coil domains.

Probably the best-characterized function of SSIII lies in the synthesis of long, cluster-spanning B chains (i.e., B_2, B_3, and so on). Alterations in the short-chain profile of amylopectin from *ssIII* mutants indicate that SSIII is also involved in the synthesis of short A and B chains. These changes are subtle when compared to those caused by the lack of SSI or SSII. Partially redundant functions between SSII and SSIII are also suggested.

3.3.3 INITIATION OF GRANULE FORMATION: STARCH SYNTHASE IV (SSIV)

3.3.3.1 Introduction

To date, mutants with reduced SSIV activity were described only in rice and *Arabidopsis*.[15] Rice has two SSIV isoforms: *Os*SSIVa, which is expressed only little in leaves and endosperm, and *Os*SSIVb, which is generally highly expressed. Neither single *repressors* of *Os*SSIVa or *Os*SSIVb nor *null mutants* of *Os*SSIVb show marked alterations in starch content or structure in the seed endosperm. In contrast, null mutants of the single SSIV isoform in *Arabidopsis* show alterations in the diurnal leaf starch content. The *Arabidopsis ssIV* mutant also has remarkable alterations in the number and shape of starch granules.

Overall, *Arabidopsis* SSIV has a unique function regarding the initiation and morphology of starch granules and the degree of starch accumulation.[6] Interestingly, the levels of ADPglucose in *ssIV* mutants are increased over 50-fold suggesting that its consumption is strongly limited. This seems to be the case in starch-free chloroplasts in which the remaining SS isoforms lack a glucan substrate. In contrast, *Arabidopsis* plants having SSIV as the sole soluble SS (i.e., *ssI/ssII/ssIII* mutants) produce similar numbers of granules per chloroplast as the wild type, despite having only little starch.

Analyses of multiple SS mutants in *Arabidopsis* indicate that SSIII, but not SSI and SSII, is required to achieve some starch synthesis in the absence of SSIV: *Arabidopsis ssIII/ssIV* mutants almost completely fail to synthesize starch. Despite a strongly reduced rate of photosynthesis, these mutants still accumulate around 100 times more ADPglucose. The *ssIIIa/ssIVb* double mutant in *japonica* rice was still able to produce substantial amounts of endosperm starch but accumulated spherical and loose granules instead of the polyhedral compound-type starch granules normally observed.

The SSIII and SSIV share the structural similarity in having a very long N-terminal extension compared with the protein structures of SSI, SSII, and GBSS.[2] A distinct functional interaction exists between these enzymes. It was found that SSIV could use maltose and maltotriose as glucan primers for glucan synthesis, and the synthetic rate with maltotriose is higher than 90% of that from amylopectin, suggesting the additional role of SSIV in glucan synthesis.

The function of SSIV and the mechanisms for determining granule number and shape are unclear. Localization studies of SSIV report it to be a stromal protein loosely targeted to distinct spots within the chloroplast—initially proposed to be the edges of granules. However, more recently, it was reported to be associated with thylakoid membranes. Given its hypothesized role in determining granule number and morphology, the specific positioning of SSIV within the plastid may well be a factor of critical importance. In turn, understanding the factors that influence its localization—to the granule or membranes—is likely to give a completely new insight into the control of starch formation.

Additional factors have been implicated in the control of starch granule number. Interestingly, mutant phenotypes have suggested the involvement of α-glucan phosphorylase and isoamylase DBEs as influencing starch granule number. It should also be noted that one of the proteins implicated in the reversible glucan phosphorylation, whose activity is required for starch degradation at night, has a profound impact on starch granule morphology, as its loss resulted in very large, round granules.

3.3.3.2 Crystal Structure of SSIV from *Arabidopsis thaliana*

In 2018, Nielsen[26] reported the crystal structure of the catalytic domain of SSIV from *Arabidopsis thaliana*, with the enzyme bound to ADP and acarbose (Figure 3.11).

The crystal of the catalytic domain of AtSSIV has two crystallographically independent SS molecules (chain A bound to ADP and acarbose and chain B bound to ADP and glucose) in their asymmetric units. Both protein molecules are, as for CLg1GBSS, in the closed conformation of the active site, bound to ADP and acarbose, and modeled without any missing loops. The SSIV structure illustrates

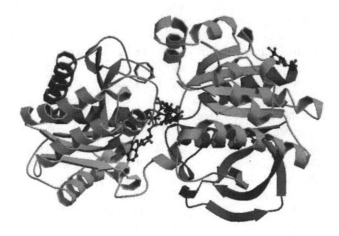

FIGURE 3.11   Catalytic domain of starch synthase IV from *Arabidopsis thaliana* bound to ADP and acarbose; RCSB PDB 6GNE. (Reproduced with permission from RCSB.)[20,29]

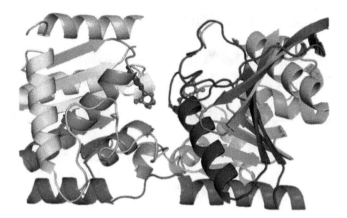

FIGURE 3.12 Structure of chain A bound to ADP, acarbose, and a surface maltose unit in the AtSSIV crystal. (Open-access article distributed under the terms of the Creative Commons Attribution License (CC BY).)[26]

the modes of binding for both donor and acceptor in a plant SS (Figure 3.12).[26] The structure is representative of the conformation of the productive ternary complex among enzyme, sugar donor, and acceptor.

3.4 BRANCHING ENZYMES

3.4.1 DOMAIN STRUCTURE

Starch-branching enzymes (E.C. 2.4.1.18) belong to the α-amylase superfamily of enzymes (also termed glycoside hydrolase family 13, GH13; CAZy).[6,30] They cleave an α-1,4-glucan chain and transfer the cleaved portion to the C6 position of a glucose unit from the same or another chain, creating an α-1,6 linked branch. In this way, BEs generate additional substrates for the SSs (i.e., non-reducing ends of chains). BEs share a common three-domain structure: an N-terminal domain containing the CBM of family 48, a central catalytic α-amylase domain characteristic for GH13 family members, and a C-terminal domain present in several α-amylases (Figure 3.13).

According to sequence similarities, BEs are separated into class I (or family B) and class II (family A) enzymes. Class I enzymes preferentially transfer longer chains than class II enzymes.[31] Genetic analysis shows that the two classes of BE make distinct contributions to the synthesis of amylopectin. Class I enzymes generally occur as a single *isoform* in plants, except for *Arabidopsis*, which does not contain a class I BE. Class II is represented by a single gene in potato and pea, but two isoforms—BEIIa and BEIIb—occur in the cereals. These have distinct expression patterns: in maize, rice, and barley, the expression of BEIIb is restricted to the grain, whereas BEIIa is found in all tissues but at often lower levels. *Arabidopsis* also has two type II BEs (BE2 and BE3), both of which are expressed in leaves and appear functionally redundant.

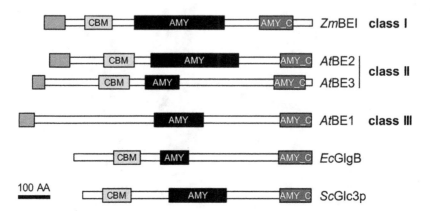

FIGURE 3.13 The domain structure of branching enzymes (BEs). BE classes can be distinguished by sequence but share a common domain structure. Depicted are plastidial transit peptides (N-terminal *blue boxes*), carbohydrate-binding modules of family 48 (*orange boxes*, CBM), catalytic α-amylase family domains (*black boxes*, AMY), and the all-β-domains typically found in the C terminus of α-amylase family members (*blue boxes*, AMY_C). The domain structure is conserved between classes and orthologs. As *Arabidopsis* does not have a class I BE, maize BEI (*Zm*BEI) is included for comparison, as are glycogen BEs from budding yeast (*Sc*Glc3p) and *E. coli* (*Ec*GlgB). *At*BE1 is a putative BE of class III which lacks the CBM48 and has not yet shown branching activity. Bar 100 amino acids (AA). (Reproduced with permission from Springer Nature.)[9]

Protein-sequence homology revealed a putative third class of BE. Genes belonging to this class III were found in *Arabidopsis* (BE1), rice, poplar, and maize (BEIII).[6] Despite the high homology within this class (~60 %), the proteins share only ~30 % identity with BEs from class I and II. To date, functional analysis on this putative class was only done in *Arabidopsis*. Branching activity was not yet reported by either group.

The specific placement of branches by BEs is believed to be a major determinant of the cluster structure of starch.[6] Class I BEs are more active toward amylose, whereas class II BEs typically prefer amylopectin. Furthermore, BEI from rice modified phosphorylase-limit amylopectin, suggesting that this BE can transfer already-branched chains.

3.4.2 Crystal Structure of Branching Enzyme I (BEI)

In 2011, Noguchi et al.[32] determined the crystal structure of the starch-branching enzyme (BEI) from *Oryza sativa* (Asian rice) (Figure 3.14).

Sequence analysis of the rice-branching enzyme I indicated a modular structure in which the central α-amylase domain is flanked on each side by the N-terminal CBM 48 and the α-amylase C-domain.[32] Noguchi et al.[32] determined the crystal structure of BEI at a resolution of 1,9 Å by molecular replacement using the *E. coli* glycogen BE as a search model. Despite three modular structures, BEI is roughly ellipsoidal with two globular domains that form a prominent groove, which is proposed to serve

FIGURE 3.14 Structure of the starch-branching enzyme I (BEI) from *Oryza sativa* L. (Asian rice); RCSB PDB 3AMK. (Reproduced with permission from RCSB.)[33]

as the α-polyglucan-binding site. Amino acid residues Asp344 and Glu399, which are postulated to play an essential role in catalysis as a nucleophile and a general acid/base, respectively, are located at a central cleft in the groove. Moreover, structural comparison revealed that in BEI, extended loop structures cause narrowing of the substrate-binding site, whereas shortened loop structures make a larger space at the corresponding subsite in the *Klebsiella pneumoniae* pullulanase. This structural difference might be attributed to distinct catalytic reactions, transglycosylation and hydrolysis, respectively, by BEI and pullulanase.

In 2012, Chaen et al.[34] determined the crystal structure of the starch-branching enzyme I complexed with maltopentaose from *Oryza sativa* L. (Figure 3.15).

Maltopentaose is bound to a hydrophobic pocket formed by the N-terminal helix, CBM48, and α-amylase domain.[34] In addition, glucose moieties could be observed

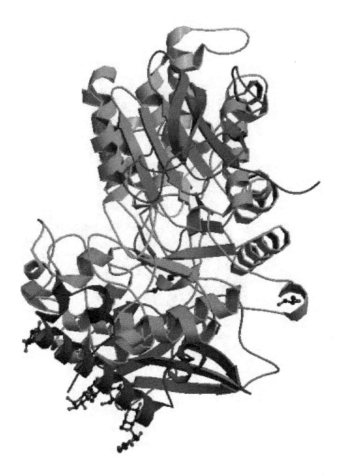

FIGURE 3.15 Structure of the starch-branching enzyme I (BEI) complexed with maltopen-taose from *Oryza sativa* L. (Asian rice); RCSB PDB 3VU2. (Reproduced with permission from RCSB.)[35]

at molecular surfaces on the N-terminal helix and CBM48. Amino acid residues involved in the carbohydrate bindings are highly conserved in other SBEs, suggesting their generally conserved role in substrate binding for SBEs.

3.5 DEBRANCHING ENZYMES

3.5.1 DOMAIN STRUCTURE

Plant DBEs hydrolyze α-1,6-linkages and release linear chains. They belong to the glycoside hydrolase family 13, GH13 (CAZY[36]), and share the central α-amylase domain and a starch-binding domain with BEs (Figure 3.16).[6] They can be further divided into two types: isoamylases (ISAs; E.C. 3.2.1.68) and limit-dextrinase (LDA; E.C. 3.2.1.41). The two types can be distinguished by protein sequences and substrate specificity, for example only LDA can degrade *pullulan*, a yeast-derived glucan

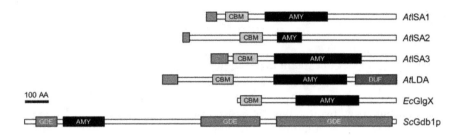

FIGURE 3.16 The domain structure of debranching enzymes.[9] The *Arabidopsis* debranching enzymes (*At*ISA1, *At*ISA2, *At*ISA3, and *At*LDA) and *E. coli* debranching enzyme (*Ec*GlgX) share the domain structure with BEs. The indirect debranching enzyme from budding yeast (*Sc*Gdb1p) is shown for comparison. Depicted are the plastidial transit peptides (N-terminal *boxes*), carbohydrate-binding modules of family 48 (CBM), catalytic α-amylase family domains (AMY), and a Pfam domain (protein families) of unknown function (DUF). Where full-length sequences could be obtained, the domain structures of classes were conserved among orthologs in various species. *Sc*Gdb1p contains three Pfam glycogen-debranching enzyme (GDE) domains. Bar 100 amino acids (AA). Reproduced with permission from Springer Nature.

consisting of α-1,6-linked maltotriosyl units (hence, the enzyme is also termed as pullulanase). Plant genomes encode three classes of isoamylase—ISA1, ISA2, and ISA3—and one LDA.[11]

ISA1 and ISA2 are involved in debranching during the synthesis of amylopectin.[16] They act together as a heteromultimeric enzyme, and ISA1 additionally forms active homomultimers in some species. "ISA" is used to refer to both homomeric and heteromeric activities. In contrast, ISA3 and LDA primarily debranch starch during its degradation.

3.5.2 CRYSTAL STRUCTURE

ISA1 and ISA2 are known to exist in a large complex and are involved in the biosynthesis and crystallization of starch.[37] It is suggested that the function of the complex is to remove misplaced branches of growing amylopectin molecules, which would otherwise prevent the association and crystallization of adjacent linear chains. The function of ISA1 and ISA2 from starch producing alga *C. reinhardtii* has been investigated. Analysis of the homodimeric CrISA1 structure reveals a unique elongated structure with monomers connected end-to-end. The crystal complex reveals details about the mechanism of branch binding that explains the low activity of CrISA1 toward tightly spaced branches and reveals the presence of additional secondary surface carbohydrate-binding sites.

The overall structure of CrISA1 is composed of three domains, characteristic of members of family 13 glycoside hydrolase (GH13) subfamily 11 (CAZy).[37,38] These include the highly conserved catalytic (β/α)$_8$ A-domain (aa 184–750), N-terminal β-sandwich domain (aa 76–183), and C-terminal β-sandwich domain (aa 751–875).

Mutants lacking ISA partly replace starch by a water-soluble polysaccharide in the endosperms of maize, rice, and barley, in potato tubers, *Arabidopsis* leaves, and

C. reinhardtii.[6] This water-soluble glucan is suggestive of glycogen and was hence called phytoglycogen. In most cases, insoluble starch, albeit with small structural alterations, is still made.

A common model to explain the accumulation of *phytoglycogen* is the "trimming" model.[6] According to this model, ISA removes excess branches from a newly created amorphous zone in "pre-amylopectin" so that only the appropriate chains are elongated by SSs. In the absence of ISA, the high number of branches would result in steric limitations, abolishing regular structures and the synthesis of further lamellae.

3.6   FACTORS POSSIBLY INFLUENCING STARCH SYNTHESIS

Evidence has been obtained for the formation of complexes between starch-biosynthetic enzymes and the phosphorylation of these enzymes.[6] Recently, two additional complexes were identified, which are distinct as they involve proteins without apparent catalytic functions. It was observed that BEIIb interacts with BEI and with starch phosphorylase in the wheat endosperm. Phosphorylation was proposed to be a prerequisite for complex formation, consistent with the observation that all wheat BEs can be phosphorylated.

A potential biological function for complex formation could be to channel substrates from one enzyme to another—a newly created branch resulting from BE action could be directly elongated first by SSI and then by SSII, thereby increasing the overall efficiency of the process. It is also possible that the complex confers enzymatic specificity; for instance, steric limitations of the whole complex could define the length of a chain that is transferred by a BE or where exactly a new branch is placed. Furthermore, it was suggested that complex formation could regulate enzyme activity allosterically and/or could protect a growing glucan from degrading enzymes. Still, none of these hypotheses has yet been tested. There is a need for detailed biochemical and structural analyses of these protein complexes in the future.

The reversible phosphorylation of starch is best characterized in *Arabidopsis*, where it is crucial for efficient degradation.[6] It is believed that phosphorylation of glucosyl units at the C6 or C3 positions disrupts and destabilizes the helical structures formed by glucan chains, rendering them accessible to degrading enzymes, in particular β-amylases.

α-Glucan phosphorylase (EC 2.4.1.1) catalyzes the reversible reaction described in the following Equation (3.4):

$$(\alpha\text{-}1,4\text{-glucose})_n + \text{G1P} \rightleftarrows (\alpha\text{-}1,4\text{-glucose})_{n+1} + P_i \tag{3.4}$$

Higher plants have two classes of phosphorylase: a plastidial and a cytosolic one. The plastidial class, named Pho1, PhoL, or PHS1, carries a 78-amino acid insertion near the glucan-binding site and shows the highest affinity toward linear glucans, especially maltooligosaccharides. By contrast, the cytosolic class, named Pho2, PhoH, or PHS2, prefers branched substrates such as glycogen. It is not clear whether Pho1 acts in a synthetic or phosphorolytic (i.e., degrading) way in vivo.

Several novel proteins have been identified that influence starch structure and/or amount.[6] In some cases, the impact on starch may be an indirect one, whereas in others, the proteins may be part of an undiscovered aspect of the starch-biosynthetic process. In either case, the analysis of the underlying mechanisms has the potential to provide new insights into starch biosynthesis that go beyond the core enzyme machinery.

3.7 OVERVIEW

Starch is the main storage carbohydrate in plants. It can be divided into two types, transitory starch and storage starch.[9] The starch that is synthesized in the leaves directly from photosynthates during the day is defined as transitory starch since it is degraded in the following night to sustain metabolism, energy production, and biosynthesis in the absence of photosynthesis. If this night-time carbohydrate supply is reduced, plants grow more slowly and experience acute starvation. The starch in non-photosynthetic tissues is generally stored for longer periods and regarded as storage starch. Remobilization of starch takes place during germination, sprouting, or regrowth, again when photosynthesis cannot meet the demand for energy and carbon skeletons for biosynthesis.[9]

Recent studies have suggested that the plastidial pathway of starch synthesis exists in all extant higher plants and green algae and that the starch biosynthetic enzymes of higher plants underwent a complex sequence of changes during evolution. The biosynthesis pathway of storage starch is illustrated in Figure 3.17.[15]

Starch biosynthesis is a highly regulated metabolic process that requires the coordinated activities of multiple enzymes, and most of the enzymes involved in these catalytic reactions are the same between amyloplasts and chloroplasts.[15] In green plants, the starch biosynthesis pathway involves a complex network of genes, most of which are members of large multigene families with multiple isoforms. However, the starch synthesis-related enzyme isoforms have not yet been identified and classified in some plants. AGPase, as the first enzyme in the starch biosynthesis pathway, catalyzes the limiting reaction by converting G1P and ATP to ADPglucose and PP_i in leaves and heterotrophic tissues. It functions as the major regulatory step in the starch biosynthetic process.[39] In most tissues, AGPase is located exclusively in the plastid. In leaves in the light, G1P is synthesized from Calvin–Benson cycle intermediates, while ATP is provided by photophosphorylation at the thylakoid membrane and imported into the plastid. In non-photosynthetic tissues, incoming sucrose is mobilized by a series of cytosolic reactions to G6P, which is imported into the amyloplast and then converted to G1P via plastidial phosphoglucomutase (PGM).

SS can be divided into GBSS, which is responsible for the synthesis of amylose and the extra-long-chain fraction of amylopectin and soluble SSs, which are mainly responsible for the synthesis of amylopectin.

SBEs belong to the α-amylase family. They cleave an α-1,4-glucan chain and transfer the cleaved portion to the C6 position of a glucose unit from the same or another chain, creating an α-1,6 linked branch.

DBEs are other glucan-modifying enzymes that occur in two forms, namely, isoamylase-type DBE (ISA) and pullulanase-type DBE (PUL). ISA generally acts

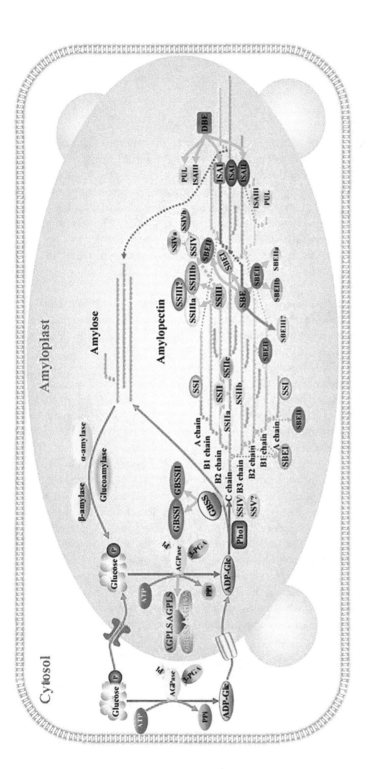

FIGURE 3.17 The storage starch biosynthesis pathway. AGPase synthesizes ADP-glucose from G1P and ATP; it is a heterotetramer (L_2S_2) consisting of two large and two small subunits; the smaller subunit plays a catalytic function, while the larger subunit is mainly responsible for modulating the regulatory properties of AGPase. Amylose is mainly produced via the activity of GBSS. Amylopectin synthesis depends on coordinated interactions among different genes encoding isoforms of SS, SBE, ISA, PUL, and PHOI. Of these, SSI plays an important role in elongating short chains to DP 8–12 in the A or B1-chains of amylopectin. SSII plays a distinct role in catalyzing the formation of intermediate chains (usually DP 13–25) of amylopectin. SSIII mainly catalyzes the synthesis of amy opectin B2 to B4 chains. SSIV plays an essential role in the initiation of starch granule formation. SBEI preferentially produces longer chains (B1 to B3), while SBEIIa and SBEIIb preferentially promote the production of short amylopectin chains (DP 6–12). The ISAI homomultimer and/or ISAI/ISAII heteromultimer have a higher affinity for relatively long external branches, while ISAII recognizes special branch points and facilitates the ability of ISAI to remove nearby branches. Additionally, ISAIII plays a major role in the starch breakdown by debranching short external chains. PUL has partially overlapping functions with ISA and is involved in cleaving short branched chains. (Courtesy of Scientific Reports.)[15]

upon amylopectin by hydrolyzing the α-1,6-linkages of polyglucans so that only the appropriate chains are elongated by SSs. On the contrary, PUL usually cleaves the α-1,6-linkages of polyglucans in pullulan and, to a lesser degree, amylopectin.

REFERENCES

1. J. Preiss, K. Ball, B. Smith-White, A. Iglesias, G. Kakefuda and L. Li, Society/Host Colloquium organized and edited by P. John, 638th Meeting held at Reading University, 1991 in https://pdfs.semanticscholar.org/b194/510b399c1cb08e426d0b63ae456d4d 9af738.pdf
2. C. Martin and A.M. Smith, *The Plant Cell*, 7, 971, 1995 in http://www.plantcell.org/content/plantcell/7/7/971.full.pdf
3. Y. Nakamura Ed., Starch, Metabolism and Structure, in *Biosynthesis of Reserve Starch*, p. 161, Springer, 2015; https://www.springer.com/us/book/9784431554943 and https://books.google.be/books?id=abwlCQAAQBAJ&pg=PA163&lpg=PA163&dq=Schematic+representation+of+the+starch+biosynthetic+pathway+and+the+related+metabolism+in+photosynthetic+(a)+and+non-photosynthetic+tissues+(b).&source=bl&ots=oUtx814b_x&sig=ACfU3U30syAwQ98V811GqhhFR3Bul_cd4g&hl=fr&sa=X&ved=2ahUKEwjPxM3UkoHpAhXB2qQKHRN9D4AQ6AEwAHoECAkQAQ#v=onepage&q=Schematic%20representation%20of%20the%20starch%20biosynthetic%20pathway%20and%20the%20related%20metabolism%20in%20photosynthetic%20(a)%20and%20non-photosynthetic%20tissues%20(b).&f=false
4. H. Ashihara, A Crozier and A. Komamine Eds., *Plant Metabolism and Biotechnology*, p 7, John Wiley & Sons, 2011 in https://onlinelibrary.wiley.com/doi/book/10.1002/9781119991311
5. M.L. Kuhn, C.M. Figueroa, A.A. Iglesias, M.A. Ballicora, BMC, *Evol. Biol.*, 2013, doi: 10.1186%2F1471-2148-13-51 in https://www.ncbi.nlm.nih.gov/pmc/articles/PMC3585822/
6. M.A. Ballicora, A.A. Iglesias and J. Preiss, *Microbiol. Mol. Biol. Rev.* 67, 213, 2003 in https://www.ncbi.nlm.nih.gov/pmc/articles/PMC156471/.
7. U.S. National Library of Medicine in https://pubchem.ncbi.nlm.nih.gov/compound/Adenosine-diphosphate-glucose#section=2D-Structure
8. R.J. Mason-Gamer, C.F. Weil and E.A. Kelloqq, *Mol. Biol. Evol.*, 15, 1658, 1998 in https://www.ncbi.nlm.nih.gov/pubmed/9866201
9. B. Pfister and S.C. Zeeman, *Cell. Mol. Life Sci.*, 73: 2781, 2016 in https://www.ncbi.nlm.nih.gov/pmc/articles/PMC4919380/
10. J.M. Berg, J.L. Tymoczko and L. Stryer, *Biochemistry*, 5th edition, W.H. Freeman, 2002 in https://www.ncbi.nlm.nih.gov/books/NBK21154/
11. Westyern Oregon University, *Sucrose & Starch Synthesis* in http://www.wou.edu/~guralnl/gural/330Sucrose%20&%20Starch%20synthesis.pdf
12. *Calvin Cycle* in https://simple.wikipedia.org/wiki/Calvin_cycle
13. Oxford Academic, *J. Exper. Bot.*, 67, 4067, 2016 in https://doi.org/10.1093/jxb/erv484
14. R.E. Koning, *Calvin Cycle*, 1994 in http://plantphys.info/plant_physiology/calvincycle.shtml
15. J. Qu, S. Xu, Z. Zhang, G. Chen, Y. Zhong, L. Liu, R. Zhang, J. Xue and D. Guo, Pmc, US National Library of Medicine, National Institutes of Health, *Sci. Rep.*, 8, 12736, 2018 in https://www.ncbi.nlm.nih.gov/pmc/articles/PMC6109180/
16. Merida, J.M. Rodriguez-Galan, C. Vincent and J.M. Romero, *Plant Physiol.*, 120, 401, 1999 in https://www.ncbi.nlm.nih.gov/pmc/articles/PMC59278/
17. M. Sjoo and L. Nilsson (Eds), *Starch in Food*, 2nd edition, Elsevier, 2017 in https://www.elsevier.com/books/starch-in-food/sjoo/978-0-08-100868-3

18. X. Jin, M.A. Ballicora, J. Preiss and J.H. Geiger, *Embo J.*, 24, 694, 2005 and PDB, 2005 in https://www.embopress.org/doi/10.1038/sj.emboj.7600551, https://www.rcsb.org/structure/1YP4 and https://www.ncbi.nlm.nih.gov/Structure/pdb/1YP4

19. K. Sarma et al., *Biomed Res Int.*, 583606, 2014 in https://www.ncbi.nlm.nih.gov/pmc/articles/PMC4167649/

20. Tuncel, I.H. Kavakli and O. Keskin, *Biophys. J.*, 95, 3628, 2008 in https://www.ncbi.nlm.nih.gov/pmc/articles/PMC2553129/

21. M.L. Kuhn, C.A. Falaschetti and M.A Ballicora, *J. Biol. Chem.*, 284, 34092, 2009 in https://www.ncbi.nlm.nih.gov/pmc/articles/PMC2797180/

22. University of Oklahoma, Department of Chemistry & Biochemistry, *Symmetry in Crystallography*, 2019 in http://xrayweb.chem.ou.edu/notes/symmetry.html

23. Wikipedia, *Tetrahedral Symmetry* in https://en.wikipedia.org/wiki/Tetrahedral_symmetry

24. CArbohydrate-active enZYmes, GT5 in http://www.cazy.org/GT5.html

25. K. Furukawa, M. Tagava, K. Tanizawa, T. Fukui, *J. Biol. Chem.*, 268, 23837, 1993 in http://www.jbc.org/content/268/32/23837.full.pdf

26. M.M. Nielsen, C. Ruzanski, K. Krucewicz, A. Striebeck, U. Cenci, S.G. Ball, M.M. Palcic and J.A. Cuesta-Seijo, *Front. Plant Sci.*, doi:10.3389/fpls.2018.01138, 2018, in https://www.frontiersin.org/articles/10.3389/fpls.2018.01138/full

27. J.A. Cuesta-Seijo, M.M. Nielsen, L. Marri, H. Tanaka, S.R. Beeren and M.M. Palcic, *Acta Crystallogr. D. Biol. Crystallogr.*, 69, 1013, 2013 in https://www.ncbi.nlm.nih.gov/pubmed/23695246

28. RCSB PDB, 4HLN, *Structure of Barley Starch Synthase I in Complex with Maltooligosaccharide* in https://www.rcsb.org/structure/4HLN

29. RCSB PDB, 6GNE, *Catalytic Domain of Starch Synthase IV from Arabidopsis thaliana Bound to ADP and Acarbose* in https://www.rcsb.org/structure/6GNE

30. CArbohydrate-active enZYmes, GH13 in http://www.cazy.org/GH13.html

31. S.C. Zeeman, J. Kossmann and A.M. Smith, *Annu. Rev. Plant Biol.* 61, 209, 2010 in https://www.researchgate.net/publication/41654872_Starch_Its_Metabolism_Evolution_and_Biotechnological_Modification_in_Plants

32. J. Noguchi, K. Chaen, N.T. Vu, T. Akasaka, H. Shimada, T. Nakashima, A. Nishi, H. Satoh, T. Omori, Y. Kakuta, and M. Kimura, *Glycobiology*, 21, 1108, 2011 in https://academic.oup.com/glycob/article/21/8/1108/1988576 and https://www.ncbi.nlm.nih.gov/Structure/pdb/3AMK

33. RCSB PDB, 3AMK in https://www.rcsb.org/structure/3AMK

34. K. Chaen, J. Noguchi, T. Omori, Y. Kakuta and M. Kimura, *Biochem. Biophys. Res. Commun.*, 424, 508, 2012 in https://www.ncbi.nlm.nih.gov/pubmed/?term=22771800

35. RCSB PDB: 3VU2 in https://www.rcsb.org/structure/3vu2

36. CAZY in http://www.cazy.org/GH13_characterized.html

37. L. Sim, S.R. Beeren, J. Findinier, D. Dauvillee, S.G. Ball, A. Henriksen and M.M. Palcic, *J. Biol. Chem.* 289, 22991, 2017 in https://www.ncbi.nlm.nih.gov/pmc/articles/PMC4132799/

38. B.L. Cantarel, P.M. Coutinho, C. Rancurel, T. Bernard, V. Lombard and B. Henrissat, *Nucleic Acids Res.*, 37, D233, 2009 in https://www.ncbi.nlm.nih.gov/pubmed/18838391/

39. A. Bahaji, J. Li, A.M. Sanchez-Lopez, E. Baroja-Fernandez, F.J. Munos, M. Ovecka, G. Amalgro, M. Montero, I. Ezquer, E. Etxeberriac and J. Pozueta-Romero, *Biotechnol. Adv.*, 32, 87, 2014 in https://crec.ifas.ufl.edu/media/crecifasufledu/faculty/etxeberria/2014-BioAdv.pdf

4 Biodegradation of Starch

4.1 INTRODUCTION

In plants (*autotrophic eukaryotes*), starch is the almost ubiquitous storage carbohydrate, whereas most *heterotrophic prokaryotes* and heterotrophic *eukaryotes* (animals, fungi) rely on *glycogen* (Figure 4.1).[1]

A common feature of photosynthetic cells is the capacity to withdraw a fraction of the carbon fixed from the Calvin cycle and thus to accumulate storage carbohydrates.[3] Carbon dioxide fixation by land plants results in production of biomass of about two hundred billion tons per year.

In leaves, transitory starch is formed in the chloroplasts during the day and broken down at night.[4] Upon degradation of starch at night, the reduced carbon stored is converted back into a metabolically active state, which can be used by many pathways of the plant. Transitory starch acts as (1) an energy reserve, providing the plant with carbohydrate during the night when sugars cannot be made by photosynthesis, and (2) an overflow when sucrose synthesis cannot keep up with rates of the rest of photosynthesis during the day, thereby allowing photosynthesis to go faster than sucrose synthesis.[5] Degradation of transitory starch mainly results in the formation of neutral sugars, such as maltose and glucose, which are transported into the cytosol.

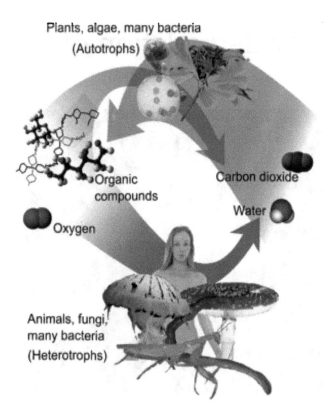

Plants, algae, many bacteria
(Autotrophs)

Organic
compounds

Carbon dioxide

Water

Oxygen

Animals, fungi,
many bacteria
(Heterotrophs)

FIGURE 4.1 Cycle between autotrophs and heterotrophs. Autotrophs use light, carbon dioxide, and water to form oxygen and organic compounds, mainly through photosynthesis. Both types of organisms use such compounds via cellular respiration to both generate ATP and again form carbon dioxide and water.[2]

Among the polysaccharides, starch and glycogen (Figure 4.2)[6] are the most widespread storage molecules.[1] Most heterotrophic prokaryotes and eukaryotes rely on glycogen for energy, whereas almost all plants form and degrade starch.

Despite obvious similarities, starch and glycogen differ in important physicochemical and biochemical properties. Glycogen is a homogeneous water-soluble polysaccharide whose branching points are uniformly distributed within the polymer. The glycogen polymer can be broken down to yield glucose molecules when energy is needed.[8] Most of the glucose residues in glycogen are linked by α-1,4-glycosidic bonds. Branches at about every tenth residue are created by α-1,6-glycosidic bonds. Recall that α-glycosidic linkages form open helical polymers, whereas β linkages produce nearly straight strands that form structural fibrils, as in cellulose.

Owing to the frequency and even distribution of the branching points, glycogen molecules are size limited. Glycogen molecules form tiny particles of 40–60 nm.[9] By contrast, starch is a water-insoluble particle without any obvious size limitation. Depending on the plant species and organ, the maximum diameter of starch granules ranges from less than 1 to 100 μm.

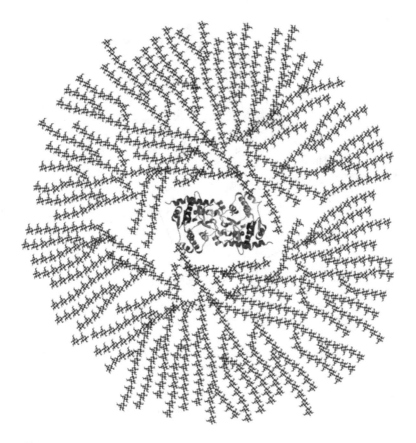

FIGURE 4.2 Schematic two-dimensional cross-sectional view of glycogen: the *glycogenin* enzyme is surrounded by branches of glucose units.[7]

4.2 TRANSITORY STARCH TURNOVER IN *ARABIDOPSIS*

4.2.1 HYDROLYTIC AND PHOSPHOROLYTIC PATHWAYS: CURRENT OVERVIEW

In leaves, transitory starch is formed in the chloroplasts during the day and degraded at night. It can be broken down hydrolytically and phosphorolytically. It is likely that the first steps of starch degradation are shared between hydrolytic and phosphorolytic pathways.[4]

The scheme described by Smith et al. in 2005 for starch degradation is presented in Figure 4.3[10] and that described by Weise et al. in 2006 is presented in Figure 4.4.[4]

To summarize the biodegradation scheme from Smith et al.,[10] glucans derived from starch granules are hydrolyzed via β-amylase (BAM) to maltose, which is exported from the chloroplast (see Figure 4.3). In the cytosol, maltose is the substrate for a transglucosylation reaction, producing glucose and a glucosylated acceptor molecule. Glucose is then converted to hexose phosphate, which leads to sucrose.[10]

In the Weise[4] scheme, the product of the phosphorolytic pathway is G1P (glucose-1-phosphate), which is formed by the action of glucan phosphorylase 1 (PHS1).

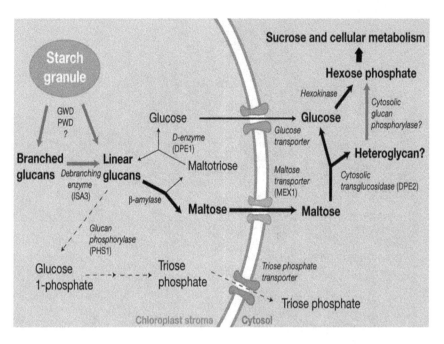

FIGURE 4.3 Proposed pathway of starch degradation in *Arabidopsis* leaves at night from Smith et al. in 2005.[10] Steps about which uncertainty remains are represented as stippled or dashed arrows and with question marks. GWD is glucan, water dikinase and PWD is phosphoglucan, water dikinase. Further details of the enzymes and reactions involved are provided in Table 4.1. (Reproduced with permission from Annual Reviews.)

The plastidic phosphorylase cannot use the intact starch granule as a substrate and prefers maltooligosaccharides to larger, branched substrates. G1P is then normally converted to G6P (glucose-6-phosphate). The products of the hydrolytic pathway, maltose and glucose, are produced by the action of BAM and D-enzyme (DPE1, disproportionating enzyme).[11,12]

4.2.1.1   Glucan Phosphorylation as the First Step in Starch Degradation

The first step in the pathway of starch degradation inside chloroplasts must be catalyzed by an enzyme capable of metabolizing polymers at the surface of a semi-crystalline granule, rather than in a soluble form. This step is the phosphorylation of a small proportion of the glucosyl residues on amylopectin by two enzymes: glucan, water dikinase (GWD) and phosphoglucan, water dikinase (PWD).[13] GWD binds to the surface of the starch granule and introduces phosphate groups at the C6 position of glucosyl units (Equation 4.1).

$$\text{glucan} + \text{ATP} \rightleftarrows \text{glucan 6-P} + P_i + \text{AMP} \tag{4.1}$$

PWD then binds to the phosphorylated starch and introduces phosphate groups at the C3 position of another glucosyl unit (Equation 4.2).

$$\text{P-glucan} + \text{ATP} + H_2O \rightleftarrows \text{P-glucan 3-P} + P_i + \text{AMP} \tag{4.2}$$

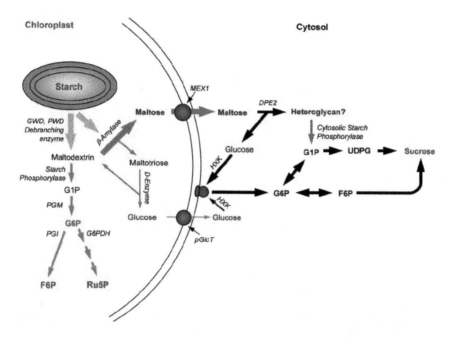

FIGURE 4.4 Proposed pathway of hydrolytic and phosphorolytic starch degradation from Weise et al. in 2006.[4] The products of the hydrolytic pathway are maltose and glucose, which are exported to the cytosol to make sucrose. The product of the phosphorolytic pathway is G1P (glucose-1-phosphate), converted to G6P (glucose-6-phosphate), which is used by the pentose phosphate pathway at night and may be used to regenerate Calvin cycle intermediates during the day. Hatched arrows indicate steps that are uncertain. GWD, glucan, water diki-nase; HXK, hexokinase; PWD, phosphoglucan, water dikinase; PGM, phosphoglucomutase; PGI, phosphoglucoisomerase; MEX1, maltose transporter; pGlcT, glucose transporter; DPE1, disproportionating enzyme 1, chloroplastic α-glucanotransferase; DPE2, disproportionat-ing enzyme 2, cytosolic α-glucanotransferase. Further details of the enzymes and reactions involved are given in Table 4.1. (Reproduced with permission from the American Society of Plant Biologists.)

4.2.1.2 Production of Maltose, Glucose, and Glucose-1-Phosphate in Chloroplasts

Several enzymes participate in the subsequent degradation of starch, including BAM, debranching enzymes, phosphoglucan phosphatases, α-amylase, disproportionating enzyme, and starch phosphorylase.[13] This results in a network of reactions rather than a linear pathway (Figure 4.5).

4.2.1.3 Metabolism of Maltose and Glucose to Hexose Phosphate in the Cytosol

Maltose produced during starch degradation is exported to the cytosol via a spe-cific transporter rather than being metabolized inside *Arabidopsis* chloroplasts (see Figures 4.3 and 4.4).[10] In addition to maltose, β-amylolytic degradation is also expected to produce a smaller amount of maltotriose because it is unable to act on chains of

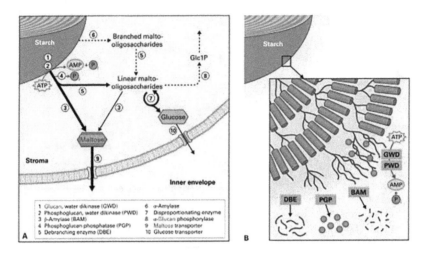

FIGURE 4.5 Pathway of starch degradation in chloroplasts. Maltose and maltooligosaccharides are released from the surface of the insoluble starch granule during degradation. Maltooligosaccharides are metabolized in the stroma, and maltose and glucose are exported to the cytosol. Estimated fluxes are indicated by relative arrow size, with dashed arrows representing minor steps in *Arabidopsis*. The inset depicts the role of phosphorylation by glucan, water dikinase (GWD) and phosphoglucan, water dikinase (PWD) in disrupting the packing of amylopectin double helices (gray boxes). This allows the release of maltose and maltose-oligosaccharides (blue lines) by β-amylases (BAMs) and debranching enzymes (DBEs). Phosphate (red dots) is released concomitantly by phosphoglucan phosphatases (PGPs) to allow complete degradation. (Reproduced with permission from Wiley.)[13]

less than four glucosyl residues. The only known maltotriose-metabolizing enzyme encoded by the genome and predicted to be chloroplastic is the disproportionating enzyme, or α-1,4 glucanotransferase (DPE1) (Table 4.1), which can potentially convert two maltotriose molecules to one maltopentaose and one glucose. This enzyme is responsible for maltotriose metabolism during starch degradation. The maltopentaose produced by DPE1 is acted on by BAM, and the glucose is exported from the chloroplast via the glucose transporter of the inner envelope (Figures 4.3 and 4.4).

Maltose is metabolized via a transglucosylation reaction.[10] Metabolism via a transglucosidase (DPE2) (Table 4.1) is the major or sole fate of maltose exported to the cytosol during starch degradation. The free glucose released from maltose via DPE2 is converted to hexose phosphate via hexokinase. Presumably, the second glucosyl unit of maltose is transferred by DPE2 to a cytosolic carbohydrate molecule (Figure 4.6).

Arabidopsis has two isoforms of α-glycan phosphorylase (EC 2.4.1.1), one residing in the plastid and the other in the cytosol.[14] The cytosolic phosphorylase, PHS2 (Table 4.1), acts on soluble heteroglycans that constitute a part of the carbohydrate pool in a plant. PHS2 is important during carbohydrate-imbalanced conditions.

Glucan phosphatases (Table 4.1) are central to the regulation of starch and glycogen metabolism.[15] Plants contain two known glucan phosphatases, Starch EXcess4 (SEX4) and Like Sex Four2 (LSF2), which dephosphorylate starch.

The combined actions of DPE2 and cytosolic α-glucan phosphorylase convert half the carbon in maltose to G1P. The other half is released as free glucose, which,

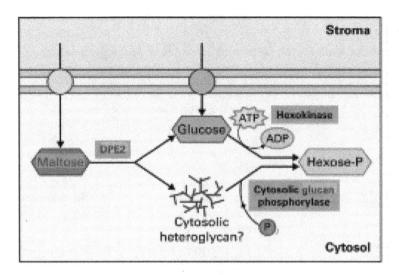

FIGURE 4.6 Maltose and glucose metabolism in the cytosol. DPE2 transfers one glucosyl unit from maltose to an acceptor molecule, probably a cytosolic heteroglucan. (Reproduced with permission from Wiley.)[14]

together with that directly exported from the chloroplast, is phosphorylated via the action of the hexokinase (Equation 4.3).

$$\text{Hexose} + \text{ATP} \rightarrow \text{hexose 6-phosphate} + \text{ADP} \qquad (4.3)$$

4.2.2 KEY ENZYMES IN STARCH DEGRADATION

As mentioned by Mahlow et al. in 2016,[16] more than 40 enzymes were discovered that are active in the process of transitory starch turnover in *Arabidopsis* (Table 4.1). The majority of starch-related enzymes are functionally located inside the plastid and have, therefore, direct access to the native starch granules. Amylolytic and phosphorolytic enzymes catalyze the hydrolysis or phosphorolysis of starch, respectively. Interestingly, none of the plastidial amylolytic enzymes act significantly on native starch granules in in vitro experiments. However, amylases considerably hydrolyze starch after disruption of the semi-crystalline structure of starch.

The simplest explanation for the involvement of GWD in starch degradation is that either the phosphate groups or the protein itself is necessary for the actions of the enzyme(s) that attack the granule surface.[10] The phosphate groups may influence the packing of the glucose polymers within the granule and hence the susceptibility of the granule surface to be attacked by enzymes. Further complexity came with the discovery that a second GWD-like enzyme, PWD (Table 4.1), is also required for normal starch metabolism. Mutants lacking this enzyme have increased amounts of leaf starch. However, the amount of phosphate in the starch of mutants lacking PWD is not dramatically affected. PWD is predicted to be chloroplastic. Although its C-terminal domain, like that of GWD, is closely related to those of other dikinases, its N-terminal domain is different from that of GWD. Recombinant PWD will phosphorylate amylopectin that already contains phosphate groups but, unlike GWD, it

TABLE 4.1

Overview of Main Starch-Degrading Enzymes in *Arabidopsis* Leaves.[10,16]

Enzyme	Reaction
Chloroplastic α-amylase, AMY3	Endoamylase that hydrolyzes α-1,4 linkages of glucans in starch. It cannot hydrolyze α-1,6 linkages. In vascular plants, α-amylases can be classified into three subfamilies.[41] *Arabidopsis* has one member of each subfamily. Among them, only *At*AMY3 is localized in the chloroplast. *At*AMY3 can act on both soluble and insoluble glucan substrates to release small linear and branched maltooligosaccharides
α-Glucan, water *dikinase*, GWD	Catalyzes the phosphorylation of starch by a mechanism in which the β-phosphate group of ATP is transferred preferentially to the C6 position of the glucosyl unit of starch.[42]
Phosphoglucan, water dikinase, PWD	Catalyzes the phosphorylation of a phosphoglucan by transferring the β-phosphate of ATP to a phosphoglucan predominantly to the C3 position of the glucosyl unit.[43]
Limit dextrinase, LDA	Hydrolyzes α-1,6 linkages of amylopectin and β-limit dextrin.
Isoamylases, ISA1 and ISA2	Hydrolyzes α-1,6 linkages of amylopectin and β-limit dextrin.
Isoamylase, ISA3	Prefers β-limit dextrin to amylopectin as a substrate.
Chloroplastic β-amylase, BAM	Exoamylase that hydrolyzes α-1,4 linkages sequentially from the non-reducing end of glucans, producing maltose. It cannot pass α-1,6 linkages.
Chloroplastic glucan phosphorylase 1, PHS1	Catalyzes the conversion of the terminal glucosyl unit at the non-reducing end to G1P, using P_i. It cannot pass α-1,6 linkages. $\text{Glucan (n)} + P_i \rightleftarrows \text{glucan (n}-1) + G1P$
Disproportionating enzyme, α-1,4- glucanotransferase, DPE1	Transfers a segment of an α-1,4-glucan (donor) to a new position in an acceptor, which may be glucose or another α-1,4-glucan.[11] This enzyme probably uses maltotriose as a substrate to transfer a maltosyl unit from one molecule to another, resulting in glucose and maltopentaose. $2\,\text{maltotriose} \rightarrow \text{maltopentaose} + \text{glucose}$
Transglucosidase, DPE2	Transfers a segment of an α-1,4-glucan (donor) to a new position in an acceptor, which may be glucose or another α-1,4-glucan.[12] It is essential for the cytosolic metabolism of maltose to hexose phosphate. It acts on the water-soluble heteroglycans (SHG) as does the phosphorylase isoform. Releases one of the glucosyl units of maltose and transfers the other to a glucan acceptor.[44,45,46] $\text{Maltose} + \text{heteroglucan acceptor (n)} \rightarrow \text{glucose} + \text{heteroglucan (n}+1)$
Cytosolic glucan phosphorylase 2, PHS2	Catalyzes the same reaction as the chloroplastic isoform, but prefers branched glucans rather than maltooligosaccharides as substrates.
Glucan phosphatases	Dephosphorylate starch and glycogen, thereby regulating energy metabolism. $\text{glucan-P} + H_2O \rightleftarrows \text{glucan} + P_i$
Hexokinase	Catalyzes the phosphorylation of hexose

will not act on unphosphorylated glucans. This suggests that PWD action in vivo requires the presence of active GWD. GWD and PWD together appear to create a pattern of phosphorylation of starch that makes it accessible to degradative enzymes at the granule surface.

The *Arabidopsis* genome encodes four proteins predicted to catalyze the hydrolysis of α-1,6 linkages in glucan polymers (Table 4.1).[10] One of these belongs to the limit dextrinase (LDA) class and the remaining three belong to the isoamylase class of debranching enzymes. LDA is present in high amounts in germinating cereal endosperm and is thought to be involved in the hydrolysis of the α-1,6 linkages in this organ. However, a mutant of *Arabidopsis* lacking LDA has normal rates of starch degradation in the leaf at night, indicating that one or more isoamylases are involved.

The initial stages of starch degradation proposed above will produce linear glucans that are soluble in the chloroplast stroma (Figure 4.3). Enzymes known to be present in the chloroplast can potentially catalyze two alternative pathways of further degradation. First, chloroplastic glucan phosphorylase (Table 4.1) can release G1P, which can be converted to triose phosphate and exported from the chloroplast in exchange for inorganic phosphate via the triose-phosphate transporter. Second, BAM can catalyze the production of maltose from linear glucans (Table 4.1). Of the nine BAM-like proteins encoded in the *Arabidopsis* genome, at least four (BAM1, -2, -3, and -4) are chloroplastic.[23] Degradation of linear glucans in the *Arabidopsis* chloroplast usually proceeds via BAMs rather than glucan phosphorylase.

4.2.2.1 Hydrolysis by Amylases

4.2.2.1.1 α-Amylases

The *Arabidopsis* genome encodes for three α-amylase-like proteins (AMY1, AMY2, and AMY3).[24] Only AMY3 has a predicted N-terminal transit peptide for chloroplastic localization. AMY3 is an unusually large α-amylase (93.5 kDa) with the C-terminal half showing similarity to other known α-amylases. AMYs and BAMs hydrolyze α-1,4-glycosidic bonds, whereas the isoamylases (ISAs) and LDAs cleave α-1,6-glycosidic bonds.[16] The action of BAMs and ISAs contributes mostly to transitory starch degradation in vivo. Mutants deficient in either BAM3 or ISA3 are impaired in starch degradation resulting in a starch excess and growth retardation phenotype.

The impact of α-amylolysis on the normal transitory starch breakdown is still not clear.[25] It has been demonstrated that α-amylase is not necessary for transitory starch degradation in *Arabidopsis* leaves.[16] *Arabidopsis* mutants deficient in all three existing α-amylases (AMY1–3) show normal starch degradation.[16] There is a major difference between starch degradation in *Arabidopsis* leaves and cereal endosperm, where starch degradation is primarily the result of α-amylolysis. There is evidence that AMY3 plays a role in leaf starch degradation in the absence of certain enzymes of the starch degradation pathway.[25] AMY3 might be responsible for the liberation of soluble, branched glucans in the absence of the DBEs ISA3 and LDA. The release of branched α-glucans by AMY3 contributes to starch degradation in DBE-deficient mutants. Furthermore, the release of glucans from native starch granules by recombinant AMY3 in vitro is stimulated by BAM.[16]

4.2.2.1.2   β-Amylases

Hydrolysis of linear α-1,4-linked glucan chains is primarily catalyzed by BAMs.[25] They liberate maltose from the non-reducing end of glucans by exoamylolysis. The *Arabidopsis* genome contains nine BAM genes (BAM1–9). Only four of these isoforms (BAM1–4) were shown to be located inside the chloroplast.[16] BAM1 is only important for starch transitory degradation in the absence of BAM3 and BAM3 plays the major role in degrading the granule. BAM4 lacks key catalytic residues present in other BAMs and loss of BAM4 results in a starch excess.

BAMs are not capable of cleaving α-1,6-linked glucose at the branch points of amylopectin.[25] Branch points are hydrolyzed by DBEs, releasing linear glucans, which are further processed by BAMs. *Arabidopsis* contains four proteins with debranching activity: ISA1, ISA2; ISA3, and LDA (also known as pullulase). LDA and ISA3 were shown to be involved in leaf starch degradation, whereas ISA1 and ISA2 have a role in the synthesis of starch. ISA3 appears to be the major DBE involved in the degradation of starch at night. The starch of the isa3 mutant is enriched in short chains. This indicates that the main role of ISA3 is to cleave α-1,6-linked glucose leading to chains degraded by BAMs.

The action of BAM will finally result in the accumulation of maltose and minor amounts of maltotriose.[25] The latter is too short to serve as a substrate for further processing by BAMs. Therefore, maltotriose is metabolized by a glucanotransferase called DPE1. Loss in DPE1 results in retardation of starch breakdown and accumulation of maltotriose. DPE1 catalyzes the conversion of two maltotriose units to glucose and maltopentaose. The latter is again used as a substrate for BAMs, while glucose is exported from the chloroplast via the plastidic glucose translocator.

4.2.2.2   Glucan Phosphorylases

During starch metabolism, the phosphorylation of glucosyl residues of amylopectin is a repeatedly observed process.[16]

4.2.2.2.1   Phosphorylation of Starch Granules

In 1998, Lorberth et al.[26] identified a non-amylolytic, starch granule-associated protein in potato, named R1, in which reduction results in a starch excess phenotype in leaves, e.g., the accumulation of high amounts of starch at the end of a normal dark phase.[16] In addition, the lowered expression of R1 is accompanied by a reduction in cold-induced sweetening in tubers. Furthermore, a reduction in the covalently attached phosphate content of starch is observed in the mutant plants. The existence of phosphoesters in starch, as the only known covalent modification of starch, was discovered in the early 1970s. The amount of starch phosphate differs according to plant species, but it is generally low. In *Arabidopsis* leaf starch and potato tuber starch, about 0.1% and 1% of the glucosyl residues are phosphorylated, respectively. Independent of the origin of starch, phosphate esters are mainly linked to the amylopectin molecule. In principle, each carbon atom of a glucosyl residue that carries a hydroxyl group can be esterified with a phosphate group. However, only G6P and G3P are found in starch hydrolysates, indicating the phosphorylation of the C6 and the C3 hydroxyl group, respectively.

Transitory starch is continuously phosphorylated during starch synthesis in the light phase, as well as during starch degradation in the dark.[16] However, the starch phosphorylation rate is higher during breakdown of starch, as revealed by radioactive labeling experiments using the unicellular green alga *Chlamydomonas reinhardtii*. Like higher plants, *C. reinhardtii* synthesizes and degrades starch in a diurnal rhythm. Moreover, starch is phosphorylated at a constant rate when the algae are illuminated. During the first 30 min of darkness, the phosphate incorporation exceeds that of illuminated cells. Likewise, phosphorylated glucans are more abundant at the granule surface of potato leaf starch during the beginning of starch mobilization.

4.2.2.2.2 Starch Phosphorylating Enzymes

The formation and the physiological function of glucan phosphorylation were unclear until Ritte et al.[27] showed in 2002 that the starch-binding R1 protein from potato catalyzes the glucan phosphorylation in a dikinase reaction type and, therefore, R1 was named glucan, water dikinase (GWD; EC 2.7.9.4).[16] As opposed to various known kinases, dikinases use ATP as a dual phosphate donor and transfer the β- and γ-phosphate groups to two distinct acceptor molecules, a glucan molecule and water.

The *Arabidopsis* mutant *sex1* (*starch excess1*) that is deficient in a protein homologous to potato tuber (*Solanum tuberosum*) GWD (StGWD) has a starch excess phenotype in leaves, which is similar to that of potato plants with reduced StGWD expression, because of impeded starch degradation and reduction in the starch phosphate content.[16] Thus, glucan phosphorylation mediated by GWD is crucial for proper starch turnover in higher plants. In in vitro assays using native starch granules, the action of recombinant GWD at the granule surface facilitates the release of maltose by plastidial BAM and ISA.

The second starch phosphorylating enzyme, PWD (AtPWD from *Arabidopsis thaliana*; EC 2.7.9.5), located in the plastids of *Arabidopsis* plants was identified later. PWD mutants have a starch excess phenotype, which, however, is not as severe as in the AtGWD mutant.

The plastidial dikinases act as monomeric proteins. They share similarities in their amino acid sequences, as well as in the catalyzed reactions. Figure 4.7 shows the schematic amino acid sequence of mature GWD and PWD including functional domains.

The highest homology between both enzymes is found in the most C-terminal region. A conserved histidine residue is capable of accepting the β-phosphate group of ATP following nucleotide binding and hydrolysis (Figure 4.5). The γ-phosphate group is transferred to water.

The largest differences in the amino acid sequence of GWD and PWD span the non-catalytic N-terminal region.[28] In the case of PWD, the N-terminus contains a single starch-binding domain (SBD) that belongs to the well-characterized CBM (carbohydrate-binding module) family CBM20. In contrast to PWD, the identity of the N-terminal starch-binding domain of GWD is less pronounced but could be assigned to the CBM45 family. In particular, two members of the CBM45 family oriented in tandem domains and separated by a 200 amino acid linker form the SBD of GWD spanning a sequence of about 500 amino acids.[29] The GWD full-length protein

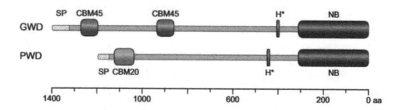

FIGURE 4.7 Schematic representation of the amino acid sequences of the two plastidial dikinases GWD and PWD from *Arabidopsis thaliania* including highlighted functional domains. The precursors of both nuclear-encoded dikinases contain an N-terminal signal peptide (SP) for plastid translocation followed by starch-binding domains. In the case of GWD, the starch-binding domain is formed by two CBM45s orientated in tandem domains, which are separated by a 200-amino acid linker. PWD has a single CBM20 starch-binding domain. The C-terminal part of both sequences shares high homology. It is characterized by a phosphohistidine domain (H*) and a nucleotide-binding domain (NB). The unit of the scale at the bottom is amino acids (aa). (Open access article distributed under the Creative Commons Attribution License (CCA 4.0).)[16]

binds to native starch granules in vivo and in vitro. Binding of GWD, as well as of PWD, in vivo is dependent on the metabolic status of the cells. A significantly higher proportion of the dikinases is associated with native leaf starch granules isolated during the dark period than in the light period.[16]

Some results indicate that GWD affects both C3 and C6 types of phosphate esters, whereas PWD is mainly responsible for C3 phosphorylation.[16] However, using labeling radioactive experiments, Ritte et al.[30] showed that GWD and PWD selectively phosphorylate the C6 and the C3 hydroxyl group of glucosyl units of starch, respectively (Figure 4.8).

4.2.2.2.3 Action of Starch-Related Dikinases on Glucan Substrates

In contrast to the plastidial amylolytic enzymes, GWD as well as PWD act significantly on the surface of native starch granules.[16] Nevertheless, the analysis of the physiological function of the dikinase-dependent starch phosphorylation and its stimulating effect on starch hydrolysis catalyzed by AtBAM3 is difficult to determine in vitro using native starch granules. Such an approach is limited because the action of both dikinases is restricted to the granule surface and glucan chains exposed at the surface account only for a minor proportion of the entire granule. Moreover, glucan chains that are phosphorylated by the dikinases remain covalently linked to the insoluble starch particle. Thus, an in vitro system was established using crystalline *maltodextrin* (MD_{cryst}) as a model substrate for glucan phosphorylating enzyme activity that mimics features of native starches, such as allomorph and crystallinity but omitted branching.[31] MD_{cryst} has a higher degree of crystallinity and, therefore, recombinant GWD phosphorylates MD_{cryst} with much higher rates than any other native starches tested so far. The incorporation of phosphate esters results in the release of phosphorylated (single, double, and triple phosphorylated) glucan chains as well as neutral maltodextrins from the water-insoluble MD_{cryst}, indicating that the action of GWD disrupts the ordered arrangement of maltodextrins at the particle surface.

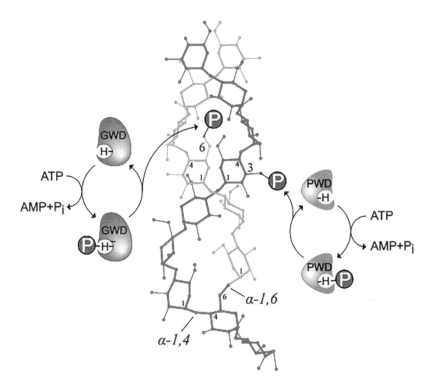

FIGURE 4.8 Schematic illustration of the plastidial dikinases-mediated reactions on α-glucan structures.[16] A left-handed α-1,4-linked glucan chain (red) forms a double helix with another glucan chain (blue) connected by an α-1,6-linkage. Both dikinases act on glucan structures and introduce phosphate esters (P); GWD (magenta) phosphorylates the hydroxyl group at carbon atom 6, whereas PWD (orange) phosphorylates the C3-hydroxyl group. Before glucan phosphorylation, both enzymes bind and hydrolyze ATP. The γ-phosphate group of ATP is transferred to water and the β-phosphate group to an auto-catalytical histidine residue (H). Subsequently, the β-phosphate is transferred to the glucan substrate. (Open access article distributed under the Creative Commons Attribution License (CCA 4.0).)

Similar to native *Arabidopsis* WT (wild-type) starch granules, significant phosphorylation of MD_{cryst} by PWD requires the preceding action of GWD.[32] However, in a strict sense, the pre-phosphorylation of GWD is not necessary for the action of PWD. It seems like that GWD-dependent phosphorylation alters the granule surface structure, which favors the action of PWD. Thus, PWD phosphorylates also non-pre-phosphorylated glucan chains. There is no relation between the endogenous starch-bound G6P content and the phosphorylation rate of PWD. The presence of small soluble maltodextrins results in a reduction of phosphorylation events on the surface of insoluble glucans. Furthermore, phosphorylation of soluble glucans is not to observe.

It can be concluded that GWD preferentially acts on crystalline surfaces and GWD-mediated phosphorylation enables a phase transition at the granule surface from a solid to a more soluble state enabling a significant amylolysis. Thus, the action of GWD is generally accepted as an initial process during starch degradation.

For normal starch degradation, the removal of the phosphate esters is necessary. Thus, the phosphate esters introduced into starch are, to a significant extent, transient and a starch phosphorylation–dephosphorylation cycle at the granule surface takes place.

4.2.2.2.4 Perspectives

Despite the knowledge acquired on dikinases, still much information is missing.[16] It is unclear whether the enzyme acts on the entire starch granule surface, what is the penetration depth of the enzyme, and whether the penetration depth is related to the phosphorylation event.

The availability of whole-genome sequences in recent years enables a view beyond higher plants and opens new perspectives in the analysis of glucan phosphorylation. In this regard, the unicellular green alga *Ostreococcus tauri* and the moss *Physcomitrella patens* are of interest. Both the genomes of *O. tauri* and *P. patens* encode for five putative starch-phosphorylating enzymes. Up to now, no data exist explaining the function of dikinase multiplicity in both organisms. The characterization of starch-related enzyme activities encoded by multiple genes in other plants, such as *Arabidopsis*, reveals distinct and, to some extent, redundant functions of enzyme isoforms. Thus, it is most likely that this holds true for the putative dikinases of *O. tauri* and *P. patens*. Elucidating the function of these putative dikinases could enlarge the toolbox of starch-modifying enzymes, which can be used in future applications to design starches with novel and desired properties.

4.2.2.3 Glucan Phosphatases

Glucan phosphatases are a recently discovered class of enzymes that dephosphorylate starch and glycogen, thereby regulating energy metabolism.[33] The general reaction catalyzed by a phosphatase is shown in Figure 4.9.[34] A phosphatase enzyme uses water to cleave a phosphoric acid monoester into a phosphate ion and an alcohol.

Glucan phosphatases are members of the protein tyrosine phosphatase (PTP) superfamily characterized by a conserved CX_5R ($CysX_5Arg$) catalytic motif (Figure 4.10).[35] The PTP superfamily consists of tyrosine-specific phosphatases, *dual-specificity phosphatases* (DSPs), and the low-molecular-weight phosphatases.[36] They are modulators of signal transduction pathways that regulate numerous cell functions. There is little sequence similarity among the three subfamilies of phosphatases. Yet, three-dimensional structural data show that they share similar conserved structural elements, namely, the phosphate-binding loop including the PTPase signature motif (H/V) $CX_5R(S/T)$ and an essential general acid/base Asp residue on a surface loop.

The characteristic sequence constitutes the phosphate-binding loop, or the so-called P-loop, where the main chain nitrogens and the guanidinium group of the

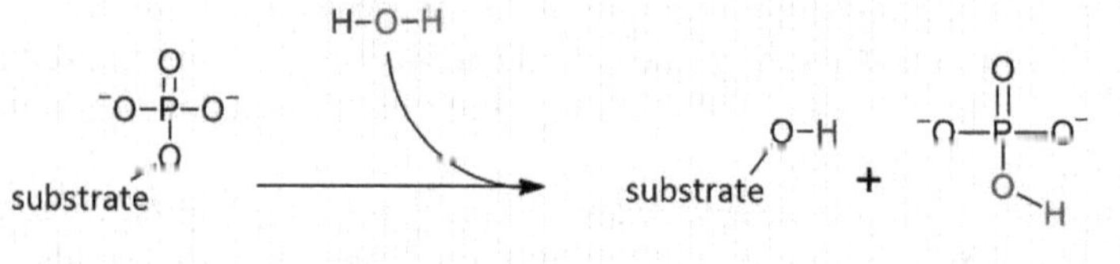

FIGURE 4.9 General reaction catalyzed by a phosphatase enzyme.[34]

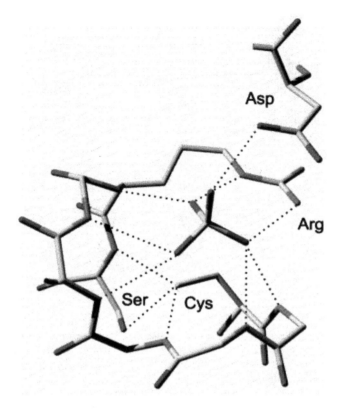

FIGURE 4.10 The structure of the characteristic sequence motif C-X$_5$-R-(S/T) and the general acid Asp in a typical protein tyrosine phosphatase. The side chains of the five residues between Cys and Arg are omitted for clarity. Hydrogen bonds stabilizing the nucleophile and the anion are indicated by dotted lines. (Reproduced with permission from Wiley.)[37]

arginine residue are oriented as to coordinate the equatorial oxygens of the phosphate group during substrate binding and catalysis. The geometry of the P-loop provides a perfect complementary structure to the two transition states of the reaction and the bidentate interaction provided by the arginine side chain is essential for catalysis.

The glucan phosphatases belong to the DSPs subfamily, with some DSPs dephosphorylating phospho-tyrosine (p-Tyr) and phospho-serine/threonine (p-Ser/Thr) residues of proteinaceous substrates, while other DSPs dephosphorylating lipids, nucleic acids, or glucans.[33,35,38]

4.2.2.3.1 Glucan Phosphatases in the Metabolism of Starch

Glucan phosphatases play a critical role in the regulation of transitory leaf starch metabolism during the diurnal photosynthesis cycle.[33] Starch permits partitioning of excess glucose produced during photosynthesis for both short- and long-term storage. The tightly packed arrangement of amylopectin helices into crystalline lamellae causes the starch granules to be semi-crystalline and water insoluble.[33]

In chloroplasts, transitory starch is continually synthesized during the photosynthetic period, and the water insolubility of starch hinders its breakdown via

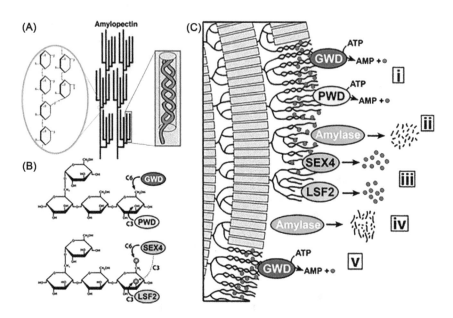

FIGURE 4.11 Reversible phosphorylation of the starch granule. (a) Amylopectin is formed from α-1,4 glucose chains with α-1,6 branches that are clustered at regular intervals (green inset). Adjacent amylopectin chains form double helices (blue inset) that expel water, contributing to the water insolubility of starch. (b) Amylopectin glucans are phosphorylated and dephosphorylated by four enzymes: GWD and PWD phosphorylate at the C6 and C3 position respectively, SEX4 dephosphorylates the C6 and C3 positions, with a preference for the C6 position, and LSF2 exclusively dephosphorylates the C3 position. (c) Reversible phosphorylation. (i) GWD and PWD phosphorylate amylopectin helices (gray bars), causing them to unwind. (ii) Amylases, particularly β-amylase, can access and degrade the glucan chains, but β-amylase is unable to hydrolyze glucans after a phosphate group (red circles). (iii) SEX4 and LSF2 dephosphorylate the exposed glucan chains, (iv) allowing further degradation. (v) The process continues with phosphorylation on the next layer of amylopectin helices within the starch granular lattice. (Reproduced with permission from Wiley.)[33]

glycolytic enzymes.[33] Starch is almost completely degraded during the night to facilitate plant growth. Therefore, during the degradative phase, hydrolytic enzymes must access the energy stored in this insoluble form. To overcome this obstacle, plants utilize a system of reversible phosphorylation that alters the biophysical properties and increases the availability of glucose chains to hydrolytic enzymes (Figure 4.11).

While the presence of phosphate in starch was first reported in the 1890s, the role of phosphate in diurnal starch metabolism first became known via biochemical experiments and the identification of mutant *Arabidopsis* plants containing a *starch excess* (SEX) phenotype.[33,39] SEX mutants have increased starch content, larger and often malformed starch granules, and show a decrease in plant growth. Together, these features indicate an inability to efficiently degrade starch granules during non-photosynthetic periods. Biochemical experiments using these mutant lines resulted in the identification of the two dikinases, GWD and PWD.

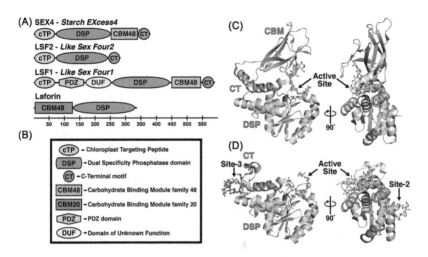

FIGURE 4.12 The glucan phosphatase family and structure of the plant glucan phosphatases. (a) Domain outline of the plant proteins SEX4, LSF2, and an inactive phosphatase LSF1, and the human protein laforin. (b) Domain legend. (c) Glucan-bound structure of **SEX4**, showing the DSP domain (blue) with the CX$_5$R catalytic site (red), the CBM (pink), and the CT motif (green). A single glucan chain (green) and phosphate (orange) were found at the active site. (d) Glucan-bound structure of **LSF2**, showing the DSP domain, CX$_5$R catalytic site, and CT motif. A single glucan chain (green) and phosphate (orange) were found at the active site. Three additional glucan chains were found at non-catalytic secondary binding sites (SBSs). A single glucan chain (cyan) was found at SBS site-2 and two glucan chains (orange and pink) were found at SBS site-3. (Reproduced with permission from Wiley.)[33]

The identification of an additional SEX mutant called *starch excess 4 (sex4)* in *Arabidopsis* resulted in the discovery of a gene encoding a trimodular protein including a chloroplast targeting peptide (cTP), a DSP domain, and a CBM (Figure 4.12a). It was subsequently demonstrated that SEX4 (Figure 4.12c) is a glucan phosphatase that dephosphorylates starch-bound phosphate incorporated by GWD and PWD. While starch phosphorylation is necessary to solubilize surface glucans, starch dephosphorylation is necessary because starch-bound phosphate groups obstruct the movement of BAM, the primary enzyme that degrades starch. B-Amylases degrade glucan chains up to a phosphate group, but SEX4 must remove the phosphate before the amylase may proceed and fully degrade the chain (Figure 4.12c). Therefore, glucan phosphatase activity essentially resets the cycle and is an essential component of efficient starch degradation. These studies demonstrated that efficient starch degradation requires the coordinated activity of dikinases, amylases, and phosphatases.

An additional glucan phosphatase called Like Sex Four2 (LSF2, Figure 4.12d) was discovered based on sequence similarity with the SEX4 DSP domain.[33] LSF2 possesses strong glucan phosphatase activity against starch. This discovery was unexpected due to the lack of a CBM in LSF2, because a CBM was predicted to be essential for glucan phosphatase activity. Furthermore, LSF2 exclusively dephosphorylates the C3-position of starch glucose units. This specificity is in contrast to SEX4, which prefers the C6-position, but can also dephosphorylate the C3-position.

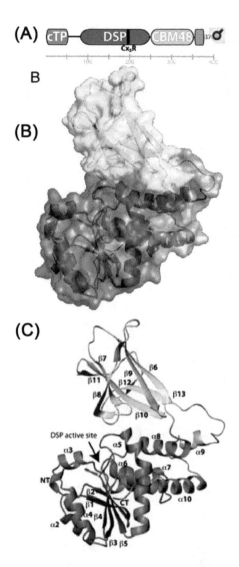

FIGURE 4.13 Structure of *At*-SEX4 determined to a resolution of 2.4 Å. (a) SEX4 domain structure. The protein includes 379 amino acids. The active site of SEX4 is denoted with a black line and labeled Cx_5R. (b) Structure of SEX4 showing the integrated architecture of DSP (pink), CBM (green), and C-terminal (blue) domains. (c) Ribbon diagram of SEX4 (residues 90–379) with the active site labeled and S198 (serine 198 phosphorylation site) in gold sticks. Elements of secondary structure are numbered from N to C termini. (Reproduced with permission from PNAS.)[40,41]

The characterization of LSF2, therefore, established an ingenious two-enzyme model for both phosphorylation and dephosphorylation of starch in plants with the activity of GWD and PWD balanced by SEX4 and LSF2.

In 2016, the X-ray crystal structures of SEX4 and LSF2 were determined, both with and without glucan ligands bound.[33] These data allowed us to establish the structural bases of their respective mechanisms, define how the glucan phosphatases incorporate phosphoglucans into their active sites, determine how the glucan chain is oriented to achieve respective substrate specificities, and hypothesize about the activity of other phosphatases.

4.2.2.3.2 Structure of SEX4 and LSF2

The first structure of a glucan phosphatase was determined in 2010 by Vander Kooi et al.[40] in the absence of a glucan. They described the crystal structure of At-SEX4, revealing its unique tertiary architecture of SEX4 (Figure 4.13).

At-SEX4 is a 379 amino acid protein containing three domains as already mentioned: an amino-terminal cTP, a DSP domain, and a CBM family member 48 (Figure 4.13a). The structure of the SEX4 glucan phosphatase reveals a unique set of extensive interdomain interactions producing a complex tertiary architecture (Figures 4.13b and 4.13c). The SEX4 DSP domain consists of a central five-stranded β-sheet flanked by eight α-helices. The SEX4 CBM domain possesses six β-sheets that fold into a characteristic compact β-sandwich composed of antiparallel sheets. The C-terminal to the CBM is an extended region that possesses two α-helices. The structure reveals an integrally folded unit composed of an N-terminal DSP domain, a CBM, and a previously unrecognized domain at the C terminus (CT).

In 2014, Meekins et al.[35] determined the structure of SEX4 bound to a phosphoglucan product, providing insights into the mechanism for C6 specificity (Figure 4.14).

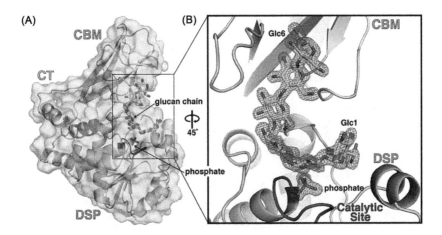

FIGURE 4.14 Crystal structure of At-SEX4 bound to a glucan chain and phosphate. (a) The surface/ribbon diagram of At-SEX4 bound to maltoheptaose (green) and phosphate (orange) is determined to a resolution of 1.65 Å. The SEX4 structure contains the DSP domain (blue) with the catalytic site (red), the CBM (pink), and the CT motif (tan). The maltoheptaose chain is located in a pocket spanning the CBM and DSP domains, and a single phosphate molecule is located at the base of the catalytic site directly beneath Glc2. (b) A close-up view rotated 45° showing the electron density map of the maltoheptaose chain (green) and phosphate (orange) bound to SEX4. The density map permitted the modeling of six glucose units in the maltoheptaose chain and the assignment of glucan orientation. Two conformers were modeled for Glc1, differing in the orientation of the O6 group. (Reproduced with permission from PNAS.)[35]

Electron density allowed the modeling of six glucose units of a maltoheptaose chain.[35] The maltoheptaose chain is located within a pocket that spans the DSP and CBM domains, with Glc1 located at the DSP and Glc6 located at the CBM. In addition, a single phosphate molecule was found within the catalytic site (PTP loop), directly below Glc2 at a distance of 2.5 Å from the catalytic residue S(C)198. The DSP–CBM pocket is ~9 Å deep and ~33 Å long with a total contact area of 610 Å². Of this contact area, 40% of the interactions occur via the CBM domain and 60% via the DSP. This study demonstrates the structural basis for SEX4phosphoglucan interactions.

Following the determination of the SEX4 structure, the structure of LSF2 was also determined (Figure 4.15).[33,42,43]

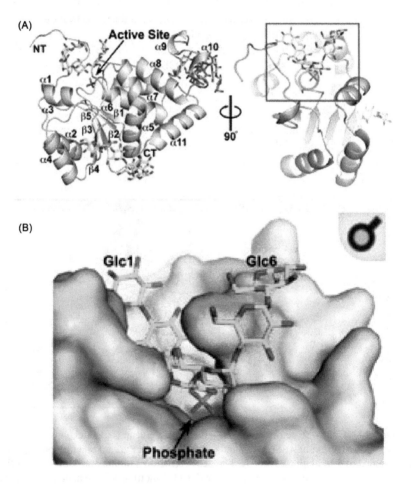

FIGURE 4.15 Structure of LSF2 bound to maltohexaose and phosphate. (a) Ribbon diagram of LSF2 (residues 79–282, C193S). Maltohexaose chains (green, cyan, orange, and pink) and phosphate (teal) are shown. Elements of secondary structure are numbered consecutively from N to C termini. (b) Maltohexaose chain (green) and phosphate (teal) at the active-site (red) channel. The image correlates with the red box in (a). Glc moieties are numbered from the non-reducing to the reducing end. (Reproduced with permission from the American Society of Plant Biologists.)[43]

LSF2 is a 32 kDa protein with 282 amino acids containing a cTP and a DSP domain (see Figure 4.12a). Sequence alignments also predicted that LSF2 contained a CT-motif similar to SEX4. The LSF2 DSP domain (residues 79–244) also possesses a characteristic αβα DSP fold consisting of a central five-stranded β-sheet region flanked by eight α-helices (see Figure 4.12d). The LSF2 CT-motif (residues 245–282) consists of a loop region culminating in an α-helix that integrally folds into the DSP domain. Although LSF2 does not contain a CBM, the structure of LSF2 bound to the glucan ligand maltohexaose revealed the presence of two non-catalytic surface-binding sites (SBSs) associated with the CT-domain in addition to a glucan ligand located at the DSP catalytic site. These results suggested that LSF2 uses these two SBSs to engage glucan ligands instead of a CBM.

4.2.2.3.4   Perspectives

Glucan phosphatases represent a new subset of PTPs that are essential for complex carbohydrate metabolism.[33] Structural and enzymatic characterization of SEX4 and LSF2 provides detailed information on the physical basis for starch interaction and dephosphorylation. SEX4 binds glucans via an extended CBM/ DSP domain interface that couples strong glucan binding at the active site with phosphoglucan integration into the catalytic site by the DSP. On the other hand, LSF2 contains SBSs that non-catalytically interact with glucan chains and a more strongly binding DSP active site that integrates the glucans into the catalytic site. A comparison of SEX4 and LSF2 with other DSPs indicates that the glucan phosphatase DSP domain contains a unique network of aromatic residues that function as glucan platforms and a wide, shallow active site to accommodate three glucan units of a longer glucan chain. Lastly, the substrate specificity of SEX4 and LSF2 is based upon discrete elements within the DSP domain and may be manipulated by simple mutagenesis in SEX4. Although SEX4, LSF2, DSP laforin, and chloroplastic LSF1 are the only known members of the glucan phosphatase family, identification of future members will most likely reveal variations on the structural patterns described.

4.3   PATHWAY OF DEGRADATION OF STARCH IN LEAVES: CURRENT VIEW

The pathway of degradation of chloroplastic starch to sucrose in leaf cells as proposed by Lloyd et al.[44] in 2016 is shown in Figure 4.16.

According to the Lloyd scheme, the steps of the chloroplastic starch degradation at night can be described as follows[44]:

- Leaf starch degradation is initiated by phosphorylation of amylopectin via glucan and water dikinases (GWD and PWD).[45]
- Dephosphorylation of the phosphoglucans by glucan phosphatases (SEX4 and LSF2) presumably occurs concurrently.
- Debranching of the starch polymers at the granule surface mainly occurs via isoamylase 3 (ISA3), and linear glucans are metabolized via BAM to yield maltose as the main product, with maltotriose as a minor product.

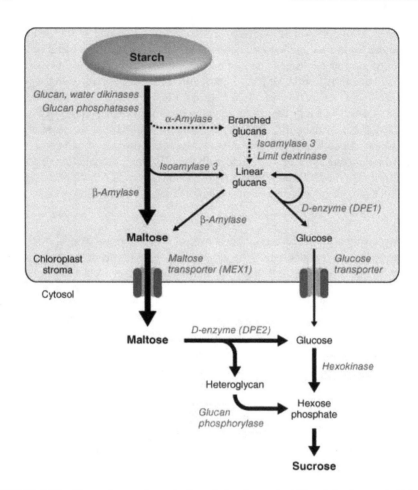

FIGURE 4.16 The pathway of starch degradation in an *Arabidopsis* leaf at night.[44] Leaf starch degradation is initiated by phosphorylation of amylopectin via glucan, water dikinase (GWD) and phosphoglucan, water dikinase (PWD). Dephosphorylation of the phosphoglucans by glucan phosphatases (SEX4 and LSF2) presumably occurs concurrently. Debranching of the starch polymers at the granule surface is mainly via isoamylase 3, and linear glucans are metabolized via β-amylase to yield maltose as the main product, with maltotriose as a more minor product. Possibly, α-amylase can release branched glucans from the granule surface, which are then debranched via the debranching enzymes isoamylase 3 and limit dextrinase. However, this is at most a minor pathway as indicated by the dashed arrows. Maltose is exported from the chloroplast to the cytosol via the maltose transporter MEX1 and then metabolized via the cytosolic disproportionating enzyme (D-enzyme; DPE2). DPE2 releases one of the glucosyl moieties of maltose as free glucose and transfers the other to a cytosolic heteroglycan, from which it is released via glucan phosphorylase as hexose phosphate. The maltotriose product of β-amylase is converted via chloroplastic D-enzyme (DPE1) to maltopentaose and free glucose. The maltopentaose is a substrate for the further action of β-amylase, and the glucose is assumed to be transported to the cytosol via a glucose transporter. For convenience, maltotriose and maltopentaose in this figure are represented under the generic term 'linear glucans'. Hexose phosphates, produced in the cytosol from free glucose and the deglucosylation of the heteroglycan, are converted to sucrose. (Reproduced with permission from Wiley.)

- Possibly, AMY can release branched glucans from the granule surface, which are then debranched via ISA3 and LDA. However, this is at most a minor pathway.
- Maltose is exported from the chloroplast to the cytosol via the maltose transporter MEX1 and then metabolized via the cytosolic disproportionating enzyme (DPE2). DPE2 releases one of the glucosyl units of maltose as free glucose and transfers the other to a heteroglycan, before their conversion to hexose phosphate via a cytosolic phosphorylase.[46]
- The maltotriose product of BAM is converted via chloroplastic D-enzyme (DPE1) to maltopentaose and free glucose.
- Hexose phosphates are converted to sucrose.

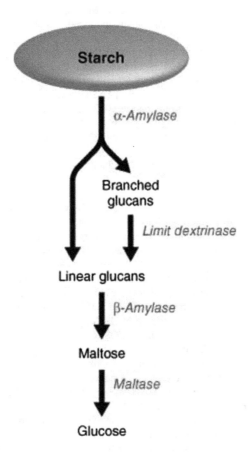

FIGURE 4.17 The pathway of starch degradation in the endosperm of a germinating cereal seed. The starch granule is attacked by the endoamylase α-amylase, which releases soluble linear and branched glucans. These are acted upon by the debranching enzyme limit dextrinase and the exoamylase β-amylase to produce maltose. Maltose is then hydrolyzed to glucose by an α-glucosidase (maltase). The glucose is taken up into the growing embryo. (Reproduced with permission from Wiley.)[44]

4.4 PATHWAY IN CEREAL SEEDS

The pathway by which starch is converted to glucose in the endosperm of germinating cereal seeds is relatively simple (Figure 4.17).[44] The starch granule is attacked by the endoamylase AMY. AMY releases both linear and branched oligosaccharides from the starch granule. These soluble products are then acted upon by two further enzymes. LDA, a debranching enzyme, generates linear chains by cleaving the α-1,6 linkages. The exoamylase BAM releases the disaccharide maltose. Maltose is hydrolyzed to two glucose molecules by an α-glucosidase (also known as maltase). All these enzymes are classified as glycoside hydrolases.

REFERENCES

1. J. Fettke, M. Hejazi, J. Smirnova, E. Hochel, M. Stage and M. Steup, *J. Exp. Bot.*, 60, 2907, 2009 in https://www.researchgate.net/publication/24237893_Eukaryotic_starch_degradation_Integration_of_plastidial_and_cytosolic_pathways
2. *Heterotroph* in https://en.wikipedia.org/wiki/Heterotroph
3. J. Smirnova, A.R. Fernie and M. Steup in Y. Nakamura (Ed.), *Starch-Metabolism and Structure*, Springer, 2015 in https://www.springer.com/us/book/9784431554943
4. S.E. Weise, S.M. Schrader, K. Kleinbeck, T.D. Sharkey, *Plant Physiol.*, 141, 879, 2006 in https://www.researchgate.net/publication/7082372_Carbon_Balance_and_Circadian_Regulation_of_Hydrolytic_and_Phosphorolytic_Breakdown_of_Transitory_Starch
5. T.D. Sharley, Commentary, Wiley, 2017, doi:10.1111/pce.13059 in https://www.osti.gov/servlets/purl/1405284
6. *Glycogen Metabolism* in http://core.ecu.edu/biol/evansc/PutnamEvans/5800pdf/Glycogen Metabolism.pdf
7. https://en.wikipedia.org/wiki/Glycogen
8. J.M. Berg, J.L. Tymoczko and L. Stryer, *Biochemistry*, 5th edition, W.H. Freeman, 2002 in https://www.ncbi.nlm.nih.gov/books/NBK21190/
9. P. Pontarotti (Ed.), *Evolutionary Biology: Convergent Evolution, Evolution of Complex Traits, Concepts and Methods*, Springer, 2016 in https://www.springer.com/us/book/9783319413235
10. A.M. Smith, S.C. Zeeman and S.M. Smith, *Annu. Rev. Plant Biol.*, 56–73, 2005 in https://www.annualreviews.org/doi/abs/10.1146/annurev.arplant.56.032604.144257 or https://www.researchgate.net/publication/7876353_Starch_Degradation
11. DPE1 in https://www.uniprot.org/uniprot/Q9LV91
12. DPE2 in https://www.uniprot.org/uniprot/Q8RXD9
13. B.B. Buchanan, W. Gruissem and R.L. Jones Eds., *Biochemistry and Molecular Biology of Plants*, Wiley-Blackwell, 2015 in https://books.google.be/books?id=F4AZCgAAQBAJ&pg=PA597&lpg=PA597&dq=dpe2+maltose+%2B+glucan+acceptor&source=bl&ots=7uxLkPR-hG&sig=ACfU3U0dl9H8rX3xjevdEX5EU8J-W0dBrA&hl=fr&sa=X&ved=2ahUKEwjLlsOE16HnAhWKalAKHf5FAjcQ6AEwAHoECAgQAQ#v=onepage&q=dpe2%20maltose%20%2B%20glucan%20acceptor&f=false
14. S. Schopper, P. Muhlenbock, S. Sorensson, L. Hellborg, *Plant Biol.*, 2014, doi: 10.1111/plb12190 in https://www.researchgate.net/publication/261992610_Arabidopsis_cytosolic_alpha-glycan_phosphorylase_PHS2_is_important_during_carbohydrate_imbalanced_conditionsJournal>
15. D.A.Meckins, M. Rathlhagala, K.D. Auger, B.D. Turner, D. Santeíia, U. Kotting, M.S. Gentry and C.W. Vander Kooi, *JBC*, 2015, M115.658203 in http://www.jbc.org/content/early/2015/07/31/jbc.M115.658203.full.pdf

16. S. Mahlow, S. Orzechowski and J. Fettke, Cell Mol. *Life Sci.*, 73 (14), 2016, doi: 10.1007/s00018-016-2248-4 in https://www.researchgate.net/publication/301832328_Starch_phosphorylation_insights_and_perspectivesJournal>
17. D. Seung, M. Thalmann, F. Sparla, M. Abou Hachem, S.K. Lee, E. Issakidis-Bourguet, B. Svensson, S.C. Zeeman and D. Santelia, *J. Biol. Chem.*, 288, 33620, 2013 in https://www.ncbi.nlm.nih.gov/pmc/articles/PMC3837109/
18. R. Mikkelsen, K.E. Mutenda, A. Mant, P. Shurmann and A. Blennow, *PNAS USA*, 102, 1785, 2005 in https://www.ncbi.nlm.nih.gov/pmc/articles/PMC547843/
19. O. Kotting, K. Pusch, A. Tiessen, P. Geigenberger, M. Steup, G. Ritte, *Plant Physiol.*, 137, 242, 2005 in https://www.ncbi.nlm.nih.gov/pmc/articles/PMC548855/
20. J. Fettke, T. Chia, N. Eckermann, A. Smith and M. Steup, *Plant J.*, 46, 668, 2006 in https://www.ncbi.nlm.nih.gov/pubmed/16640603
21. A.M. Smith, Europe PMC, *Discovery of the Pathway from Starch to Sucrose in Leaves in the Dark* in https://europepmc.org/grantfinder/grantdetails?query=pi%3A%22Smith%2BAM%22%2Bgid%3A%22BBC5073611%22%2Bga%3A%22Biotechnology%20and%20Biological%20Sciences%20Research%20Council%22
22. C. Ruzanski, J. Smirmova, M. Rejzek, R.A. Field, *J. Biol. Chem.*, 288, 28581, 2013 in https://www.ncbi.nlm.nih.gov/pmc/articles/PMC3789958/ 17.
23. D.C. Fulton, M. Stettler, T. Mettler, C.K. Vaughan, J. LI, P. Francisco, M. GIL, H. Reinhold, S. Eicke, G. Messerli, G. Dorken, K. Halliday, A.M. Smith, S.M Smith and S.C. Zeeman, *Plant Cell*, 20, 1040, 2008 in https://www.ncbi.nlm.nih.gov/pmc/articles/PMC2390740/
24. T-S Yu, S.C. Zeeman, D. Thorneycroft, D.C. Fulton, H. Dunstan, W-L Lue, B. Hegemann, S_Y. Tung, T. Umemoto, A. Chapple, D.-L. Tsai, S.-M. Wang, A.M. Smith, J. Chen and S.M. Smith, *J. Biol. Chem.*, 280, 9773, 2006 in http://www.jbc.org/content/280/11/9773.long
25. D. Fetke, PhD thesis, *The Control of Starch Degradation in Arabidopsis thaliana Leaves at Night*, University of East Anglia, Norwich, 2013 in https://ueaeprints.uea.ac.uk/48101/1/PhD_thesis_Doreen_Feike.pdf
26. R. Lorberth, G. Ritte, L. Willmitzer and J. Kossmann, *Nat Biotechnol*, 16, 473, 1998 in https://www.ncbi.nlm.nih.gov/pubmed/9592398
27. G. Ritte, J.R. Lloyd, N. Eckermann, A. Rottmann, J. Kossmann and M. Steup, *Proc. Natl. Acad. Sci. USA*, 99, 7166, 2002 in https://www.ncbi.nlm.nih.gov/pubmed/12011472
28. Brenda, *The Comprehensive Enzyme Information System* in https://www.brenda-enzymes.org/all_enzymes.php?ecno=2.7.9.5&table=General_Information
29. M.A. Glaring, M.J. Baumann, M.A. Hachem and H. Nakai, *FEBS J.*, 278, 1175, 2008 in https://www.ncbi.nlm.nih.gov/pmc/articles/PMC3516724/
30. G. Ritte, M. Heydenreich, S. Mahlow, S. Haebel, O. Kotting and M. Steup, *FEBS Lett.*, 580, 4872, 2011 in 10https://febs.onlinelibrary.wiley.com/doi/full/10.1016/j.febslet.2006.07.085
31. H. Mahdi, J. Fettke, S. Haebel and C. Edner, *Plant J.*, 55, 323, 2008 in https://www.researchgate.net/publication/5432466_Glucan_water_dikinase_phosphorylates_crystalline_maltodextrins_and_thereby_initiates_solubilization
32. H. Mahdi, J. Fettke, O. Paris and M. Steup, *Plant Physiol.*, 150, 962, 2009 in https://www.researchgate.net/publication/24361104_The_Two_Plastidial_Starch-Related_Dikinases_Sequentially_Phosphorylate_Glucosyl_Residues_at_the_Surface_of_Both_the_A-_and_B-Type_Allomorphs_of_Crystallized_Maltodextrins_But_the_Mode_of_Action_Differ
33. D.A. Meekins, C.W. Vander Kooi and M.S. Gentry, *FEBS J.*, 283, 2427, 2016 in https://www.ncbi.nlm.nih.gov/pmc/articles/PMC4935604/
34. Phosphatase in https://en.wikipedia.org/wiki/Phosphatase

35. D.A. Meekins, M. Raththagala, S. Husodo, C.J. White, H.F. Guo, O. Kotting, C.W. Vander Kooi and M.S. Gentry, *PNAS*, 111, 7272, 2014 in https://www.ncbi.nlm.nih.gov/pmc/articles/PMC4034183/

36. Z.Y. Zhang, *Crit. Rev. Biochem. Biol.*, 33, 1, 1998 in https://www.researchgate.net/publication/51334483_Protein-Tyrosine_Phosphatases_Biological_Function_Structural_Characteristics_and_Mechanism_of_Catalysis

37. K. Kolmodin, J. Aqvist, *FEBS Lett.*, 498, 208, 2001 in https://core.ac.uk/download/pdf/82812346.pdf

38. D.A. Meekins, H.F. Guo, S. Husodo, B.C. Paasch, T.M. Bridges, D. Santelia, O. Kotting, C.W. Vander KOOI and M.S. Gentry, *Plant Cell*, 25, 2302, 2013 in https://www.ncbi.nlm.nih.gov/pmc/articles/PMC3723627/Journal

39. M.S. Gentry, M.K. Brewer and C.W. Vander Kooi, *Curr. Opin. Struct. Biol.*, 40, 62, 2016 in https://www.ncbi.nlm.nih.gov/pmc/articles/PMC5161650/ and https://www.sciencedirect.com/science/article/abs/pii/S0959440X16300902?via%3Dihub

40. C.W. Vander Kooi, A.O. Taylor, R.M. Pace, D.A. Meekins, H.F. Guo, Y. Kim and M.S. Gentry, *PNAS*, 107, 15379, 2010 in https://www.ncbi.nlm.nih.gov/pmc/articles/PMC2932622/

41. RCSB, *3NME* in http://www.rcsb.org/structure/3NME

42. RCSB, *4KYQ* in https://www.rcsb.org/structure/4KYQ

43. D.A. Meekins, G. Hou-Fu, S. Husodo, B.C. Paasch, T.M. Bridges, Travis, D. Santelia, O. Kotting, C.W. Vander Kooi and M.S. Gentry, *Plant Cell*, 25, 2302, 2013 in https://www.ncbi.nlm.nih.gov/pmc/articles/PMC3723627/

44. J. R. Lloyd and O. Kotting, *Starch Biosynthesis and Degradation in Plants, eLS*, 2016, John Wiley & Sons, doi: 10.1002/9780470015902.a0020124.pub22016 in http://www.els.net/WileyCDA/ElsArticle/refId-a0020124.html and https://static1.squarespace.com/static/54694fa6e4b0eaec4530f99d/t/5bc4df58a4222fd25ae0eb23/1539628890142/Starch+Biosynthesis+and+Degradation+in+Plants.pdf

45. A.W. Skeffington, A. GRAF, Z. Duxbury, W. Gruissem and A.M. Smith, *Plant Physiol.*, 165, 866, 2014 in https://www.ncbi.nlm.nih.gov/pmc/articles/PMC4044853/pdf/866.pdf

46. C. Ruzanski, J. Smirnova, M. Rejzek, D. Cockburn, H.L. Pedersen, M. Pike, W.G.T. Willats, B. Svensson, M. Steup, O. Ebenhoh, A.M. Smith and R.A. Field, *J Biol Chem.*, 288, 28581, 2013 in http://www.jbc.org/content/288/40/28581.full

5 Properties of Starch and Modified Starches

5.1 INTRODUCTION: CORRELATION BETWEEN STRUCTURE AND PROPERTIES

Based on its biological functions, starch falls into two types: transitory starch and storage starch.[1] The starch that is synthesized in the leaves directly from photosynthates during the day is referred to as transitory starch, and it is degraded in the following night to sustain metabolism, energy production, and biosynthesis in the absence of photosynthesis. If this nighttime carbohydrate supply becomes smaller, plants grow more slowly and experience acute starvation. The starch in non-photosynthetic tissues, such as seeds, stems, roots, or tubers, is generally stored for longer periods and regarded as storage starch.

Starches from different botanical sources vary in terms of their functional properties and thus in their end uses.[1,2] This variation stems from differences in the structure of starch, such as the size of starch granules, their composition, and molecular architecture of the constituent polymers.

Extracted starch from plants often needs to be modified to confer or enhance the required functional properties. The structure of starch also influences its digestibility in the gut. Those with reduced digestibility (resistant starch), such as high-amylose starches, are increasingly valued because of their health-promoting effects, potentially serving as a preventive measure against conditions such as colorectal cancer and diabetes. Understanding starch biosynthesis and its relationships to structure and functionality is of great interest as it represents a prerequisite for the improvement of starch crops.

To enhance the properties and functionality such as solubility, texture, viscosity, and thermal stability, which are necessary for the desired product or role in the industry, native starches are modified.[3] The widening vista of application possibilities of starches with different properties has made research in non-conventional starches and other native starches more imperative. Recent studies on the relationship between the structural characteristics and functional properties of starches from different sources have continued to provide important information for optimizing industrial applications.

Modification has been achieved mostly by physical and chemical means.[3] Enzymatic and genetic modifications are biotechnological processes increasingly explored. Physical modification methods consist of superheating, dry heating, osmotic pressure treatment, multiple deep freezing and thawing, instantaneous controlled pressure-drop process, stirring ball milling, vacuum ball milling, pulsed electric fields treatment, corona electrical discharges, and so on. While these methods look generally simple and cheap, chemical modification methods involve the introduction of new functional moieties into the starch molecule via its hydroxyl groups, resulting in a significant change in its physicochemical characteristics. The functional characteristics of chemically modified starch depend on several factors including the botanical origin of the native starch, the reagent used, the concentration of reagent, pH, reaction time, the presence of a catalyst, type of substituent, degree of substitution, and the distribution of the substituents in the modified starch molecule. Chemical modification is generally achieved through chemical derivatization, such as etherification, esterification, acetylation, cationization, oxidation, hydrolysis, and crosslinking.

5.2 PHYSICOCHEMICAL PROPERTIES OF STARCH

The length of the α-glucan chains, amylose–amylopectin ratio, and branching degree of amylopectin define the size, structure, and particular utility of starch granules in each plant species.[2] Other characteristics associated with the granule such as form, surface type, and phosphate groups influence the properties and uses of starch.

5.2.1 CHARACTERISTICS OF STARCH GRANULES: MORPHOLOGY, SIZE, COMPOSITION, AND CRYSTALLINITY

Starch granules have sizes with diameters ranging from 0.1 to 200 μm, and their morphology varies between different shapes such as oval, ellipsoidal, spherical, smooth, angular, and lenticular, depending on the botanical source.[2] Size distribution can be uni-, bi-, or polymodal. In amyloplasts, starch granules are present individually or in groups. Common cereals such as wheat, barley, and rye contain two types of starch granules: A-type, lenticular shape and large size, and B-type, spherical shape and small size. Most of the native starch granules exhibit a Maltese cross.[4]

The amount of amylose present in the granule affects the physicochemical and functional properties of starch.[2] The amylose content can vary within the same botanical variety because of differences in geographic origin and cultural conditions.

Amylose plays a role in the initial resistance of granules to swelling and solubility, as swelling proceeds rapidly after leaching of amylose molecules. The capacity of amylose molecules of forming lipid complexes prevents their leaching and so the swelling capacity. Amylose is anhydrous and can form excellent films, with important characteristics for industrial applications. Films formed from amylose are very strong, colorless, odorless, and tasteless.

Phosphorus is one of the non-carbohydrate components present in the starch granule and affects its functional characteristics.[2] Phosphorus is present as monoester phosphates or phospholipids in various types of starches. Monoester phosphates are associated with the amylopectin fraction by covalent bonds, increasing the clarity and viscosity of the paste, whereas the presence of phospholipids results in opaque and low-viscosity pastes. The phospholipid content in starch granules is proportionally related to amylose. Phospholipids tend to form complexes with amylose and long branches of amylopectin, resulting in starch granules with limited solubility. The nature of the phosphorus in starch granules influences the transmittance of the paste. Starches from wheat and rice with a high phospholipid content produce pastes with low transmittance compared with potato or corn starch pastes with fewer phospholipids. Potato starch demonstrates high transmittance because of its phosphate monoester content.

Starch granules have very complex structures. Their complexity results from variations in their composition, component structure, and differences between amorphous and crystalline regions. A higher proportion of amylopectin results in higher crystallinity.

5.2.2 BIREFRINGENCE AND ORDERING

Birefringence is the optical property of a material having a refractive index that depends on the polarization and propagation direction of light.[5] Birefringence is responsible for the phenomenon of double refraction whereby a ray of light, when incident upon a birefringent material, is split by polarization into two rays taking slightly different paths.[6] Hence, birefringence is the ability to doubly refract polarized light.[2,7]

Native starch granules are birefringent when viewed under polarized light, suggesting molecular orientation of some sort in the granules. Birefringence patterns (interference cross, known as a 'Maltese cross') for granular starches are consistent with a radial direction of the macromolecules, i.e., normal to the growth rings and the surface of the granule (Figure 5.1).[8–10]

Further evidence for the radial ordering of polymer chains, which is typical of spherulitic organization, is obtained from small-angle light scattering measurements.[8,10] Optical birefringence is also exhibited by acid-treated starches and spherulitic particles of short-chain amyloses crystallized from solution. Loss of birefringence on heating, indicative of disordering processes, is used for determining gelatinization temperature.

Evidence for molecular order in granular starch is also provided by calorimetry; ordered chains give rise to endothermic transitions upon heating.[8,10] Application of differential scanning calorimetry to starch has been useful in inferring differences in starch structures, changes in physical states of starch, and interactions of starch polymers with other constituents in model systems and composite food matrices.

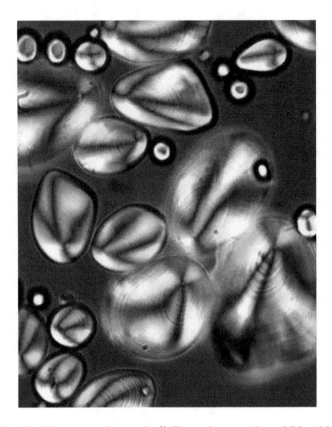

FIGURE 5.1 Birefringent starch granules.[11] The native granules exhibit a Maltese cross when observed under polarized light. (Reproduced with permission from Serge Perez.)

DSC measures the direction and extent of heat energy flow while a small sample is exposed to a constant change in temperature. It can detect both first-order (melting, crystallization) and second-order (glass) transitions of starch.

5.2.3   GLASS TRANSITION TEMPERATURE

Glass transition temperature (T_g) is an important parameter affecting the physical properties of polymers.[2] Glass transition occurs similar to a thermodynamic second-order transition, where the specific volume and enthalpy are functions of temperature. T_g describes the induction temperature of the progressive transition from an amorphous state to a rubbery state as the material is heated, generally in the presence of a solvent or plasticizer when referring to polysaccharides. Because starch consists of an amorphous and a crystalline region, the exact T_g is detected with difficulty.

The T_g of potato and wheat starches, stored for several periods after *gelatinization* (see 5.2.5), was measured by DSC, and the relative crystallinity of the starches was measured by X-ray diffractometry.[12] T_g of stored starches was higher than that of starches without storage, and the T_g increment of starches gelatinized at 120°C was higher than that of starches gelatinized at 60°C. The water content at which the glass transition of a starch occurs at 25°C was estimated at 22% from DSC data, and it increased linearly with relative crystallinity in two groups that differed in the gelatinization method. These results also showed the quantitative relationship between T_g and *retrogradation* (5.2.5). In addition, these results suggested that the glass transition of starch could be interpreted in the same way as the glass transition of crosslinked synthetic polymers.

The T_g values of starch with different amylose/amylopectin ratios were systematically studied by a high-speed DSC.[13] The corn starches with different amylose contents were used as model materials. The high heating speed (up to 300 °C/min) allows the weak T_g of starch to be visible and the true T_g was calculated by applying linear regression to the results from different heating rates. The higher the amylose content, the higher the T_g for the same kind of starch. It was found that T_g was increased from about 52°C to 60°C with increasing amylose content from 0% to 80% for the samples containing about 13% moisture.

The T_g of a biopolymer is generally affected by the polymer molar mass and the water content. As demonstrated for starch, T_g decreases with decreasing molar mass and increasing water content (Figure 5.2).[14]

In 2011, Van der Sman et al.[15] predicted the state diagram of starch–water mixtures using the Flory–Huggins free volume theory (Figure 5.3). With the estimated model parameters, they constructed the complete state diagram for starch, which can be used as a quantitative tool for the design and analysis of food structuring processes (Figure 5.4).

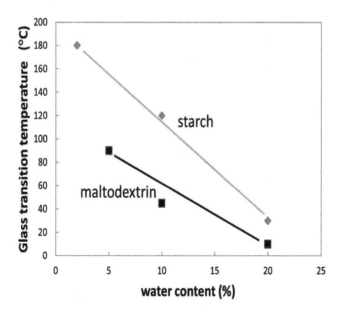

FIGURE 5.2 Effect of water content on the glass transition temperature of starch and maltodextrin with a dextrose equivalent of 20. (Reproduced with permission from the *Japan Journal of Food Engineering*.)[14]

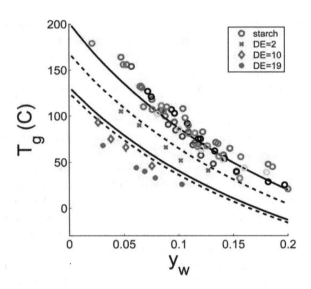

FIGURE 5.3 Glass transition temperature (T_g) of amorphous starch and different polydisperse maltodextrins (indicated with dextrose equivalent, DE) as a function of water content (y_w). (Reproduced with permission from the Royal Society of Chemistry.)[15]

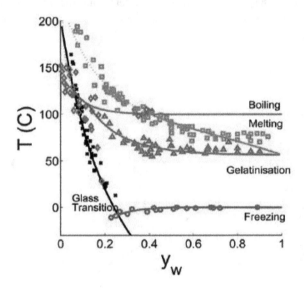

FIGURE 5.4 State diagram of starch, with some experimental data. (Reproduced with permission from the Royal Society of Chemistry.)[15]

5.2.4 SWELLING CAPACITY AND SOLUBILITY OF STARCH GRANULES

Starch granules are insoluble in cold water because of the hydrogen bonding and crystallinity of starch. When starch is dispersed in hot water below its T_g, the starch granules swell and increase several times in size, breaking the hydrogen bonding and consequently leaching amylose.

The presence of relatively short chains of amylose and amylopectin adds opacity to starch suspensions and foods containing them. In products such as dressings and puddings, this opacity is not a disadvantage; however, jellies and fruit fillings require starch suspensions with high clarity.

Starch granules are densely packed with semi-crystalline structures and have a density of about 1.5 g/cm^3.[16] Because of this stable semi-crystalline structure, starch granules are not soluble in water at room temperature. Without gelatinization, potato starch can absorb up to 0.48–0.53 g of water per gram of dry starch. The swelling process is reversible upon drying.

One of the most important structural characteristics of starch is that it passes through different stages from water absorption to granule disintegration.[2] Water absorption and consequent swelling of the starch granule contribute to amylose–amylopectin phase separation and crystallinity loss, which in turn promotes the leaching of amylose to the intergranular space. When starch molecules are heated in excess of water, the semi-crystalline structure is broken and water molecules associate by hydrogen bonding with hydroxyl groups present on amylose and amylopectin molecules. This association causes swelling and increases granule size and solubility. The swelling capacity and solubility of starch illustrate the interactions of the polymer chains in the amorphous and crystalline portions. The extent of this interaction is influenced by the amylose–amylopectin ratio and depends on the characteristics of each molecule. The swelling capacity of starch is directly linked to the amylopectin content because amylose acts as a diluent and inhibitor of swelling.

The swelling stage of starch granules is the initial step of all other paste characteristics. Initially, granule swelling is reversible, increasing its volume up to 30%.[2] Water absorption and heating of the starch dispersion break the hydrogen bonds responsible for granule cohesion, partially solubilizing the starch. Water penetrates the interior of the starch granule, hydrating the linear fragments of amylopectin. This process leads to irreversible swelling, increasing the granule size several folds and the paste viscosity. Paste viscosity is the principal measure of the potential application of starch in industry.

5.2.5 Gelatinization and Retrogradation Properties

Many food processes involve heating (cooking) and/or cooling under variable moisture conditions, which cause structural changes at the food, granular, and molecular levels.[17] Different processing conditions, therefore, have different effects on starch structure and accessibility, with implications for digestibility.

Heating native starch (50°C–100°C) in excess water results in *gelatinization*.[17] During the gelatinization process, the semi-crystalline starch granule becomes completely disrupted. Hydrogen bonds that hold the double-helical structure of α-glucan chains (the amylopectin fraction) together are broken, resulting in a greater proportion of amorphous starch material. Thus hydrothermal processing changes the morphology of starch granules, from an ordered to disordered structure (Figure 5.5). Starch gelatinization requires both heat and moisture and is most rapid when starch is heated in excess moisture (>70%) between 50°C and 100°C. In the gelatinization process and in any starch hydrothermal processing, water acts as a plasticizer.

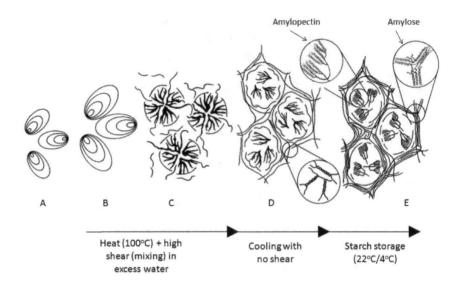

FIGURE 5.5 Starch gelatinization—effects of processing on starch granules. (a) Intact native starch granule. (b) Heat treatment in excess water under high shear conditions results in granular swelling. (c) Granule disruption occurs during starch gelatinization with linear amylose chains leaching out of the granule. (d) Upon cooling, amylose chains aggregate together to form an ordered gel network. (e) Recrystallization of amylopectin and amylose chains occurs upon storage of gelatinized starch.[17] Representation adapted from Goesaert et al. (Reproduced with permission from Elsevier.)[18]

Starches from different botanical sources, however, differ in their gelatinization behavior, and not all starch may be completely gelatinized during hydrothermal processing treatments.[17] For instance, gelatinization of high amylose starches may require temperatures above 120°C, which is considerably higher than normal.

The initiation of gelatinization is called the onset temperature. Peak temperature is the position where the endothermic reaction occurs at the maximum. Completion temperature is reached when all the starch granules are fully gelatinized and the curve remains stable. Gelatinization temperatures can be measured by DSC (Figure 5.6).[19]

Cooking a starch suspension results in rheological properties that are of value for foods. In general, starches derived from tuberous crops, such as potato or tapioca from cassava, tend to swell and thicken at lower temperatures than cereal grain starches. Cereal starches, such as maize and wheat, break down more slowly with prolonged cooking. High-amylose cereal starches are more resistant to swelling and behave like tuber starches during thermal treatments, with amylose leaching to a greater extent than amylopectin. In addition, the concentration of starch when cooked affects its swelling properties. Above certain concentrations, the swollen starch granules entrap all available water and the aqueous phase will not separate from them. Cooked granules, in contrast to unswollen ones, can be disrupted by shearing.

Gelatinization improves the availability of starch for amylase hydrolysis. So gelatinization of starch is used constantly in cooking to make the starch digestible or to thicken/bind water in roux, sauce, or soup.

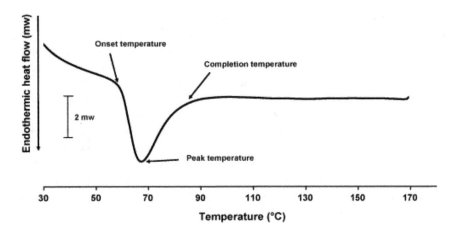

FIGURE 5.6 The gelatinization of native potato starch at 30% solid content by DSC. (Reproduced with permission from Elsevier.)[19]

Gelatinized starch, when cooled for a long enough period (hours or days), will thicken (or gel) and rearrange itself again to a more crystalline structure; this process is called *retrogradation*.[20] During cooling, starch molecules gradually aggregate to form a gel. The following molecular associations can occur: amylose–amylose, amylose–amylopectin, and amylopectin–amylopectin. A mild association among chains comes together with water still embedded in the molecule network.

Owing to strong associations of hydrogen bonding, longer amylose molecules (and starch which has a higher amylose content) will form a stiff gel. Amylopectin molecules with a longer branched structure (which makes them more similar to amylose) increase the tendency to form strong gels. High amylopectin starches will have a stable gel but will be softer than high-amylose gels. Retrogradation restricts the availability for amylase hydrolysis to occur, which reduces the digestibility of the starch.

5.2.6 RHEOLOGICAL PROPERTIES

Starch paste forms immediately after gelatinization, and starch granules are increasingly susceptible to disintegration by shearing.[2] The paste obtained is a viscous mass consisting of one continuous phase of solubilized amylose and/or amylopectin and one discontinuous phase of the remaining granules. Starch functionality is directly related to gelatinization and the properties of the paste. All these properties affect the stability of products, consumer acceptance, and production reliability. The characteristics of the native starch, the effects of the physical or chemical modifications of the granules, the process parameters, and the botanical sources of the starch are critical factors governing the behavior and characteristics of the starch paste.

Rheological properties describe the behavior of materials subjected to shearing forces and deformation, which are considered viscoelastic materials. The basic feature of starch rheology is its viscosity. Other rheological features include texture, transparency or clarity, shear strength, and the tendency for retrogradation. All these features play important roles in the commercial applications of starch. Rheological starch properties are studied through the behavior of viscosity curves.

Starch gels are composed of amylose chains and intermediate materials dispersed in a starch suspension after granule disintegration. The level and nature of the leached material and molecular interactions determine the viscoelastic properties.

The main factors affecting the rheological properties of starches are their sources and the presence of other polymers. Many polymers coexist with starch in aqueous mixtures and interact in different ways to produce attributes influencing the stability, texture, and quality of food products. Starch paste viscosity is associated with lipids, mainly phospholipids, that complex with amylose and hinder or reduce the swelling capacity of granules. Rheology is widely recognized for its effect on the quality of food and its sensory characteristics. The rheological properties of starch determine its potential applications as a thickener or gelling agent.

5.3 STARCH DIGESTIBILITY

Starch is quantitatively the major source of energy in the human diet.[2] Starch digestibility is attributed to the interaction of several factors, including the vegetal source, granule size, amylose/amylopectin ratio, degree of molecular association between components, degree of crystallinity, amylose chain length, and presence of amylose–lipid complexes on starch granules.

Starch digestion begins in the mouth, with α-amylase present in saliva and is complete when the polysaccharide is broken down into single sugars, which can be absorbed by the body (Figure 5.7).[21] Cellulose present in food is not digested by the digestive enzymes.[22]

The digestion of polysaccharides such as starch includes four steps[22]:

1. In the mouth: Digestion begins in the mouth. The salivary glands in the mouth secrete saliva, which helps moisten the food. The food is then chewed while the salivary glands also release α-amylase, which begins the process of breaking down the polysaccharides.
2. In the stomach: After the carbohydrate food is chewed into smaller pieces and mixed with salivary amylase and other salivary juices, it is swallowed and passed through the esophagus. The mixture enters then the stomach. By the time food is ready to leave the stomach, it has been processed into a thick liquid called chyme. There is no further digestion in the stomach, as the latter produces acid, which destroys bacteria in the food and stops the action of the salivary amylase.
3. In the pancreas and small intestine: After being in the stomach, the chyme enters the beginning portion of the small intestine, called the duodenum. In response to chyme being in the duodenum, the pancreas releases the enzyme pancreatic amylase, which breaks the polysaccharide down into a disaccharide. The small intestine then produces enzymes called lactase, sucrase, and maltase, which break down the disaccharides into monosaccharides. The monosaccharides are then absorbed in the small intestine.

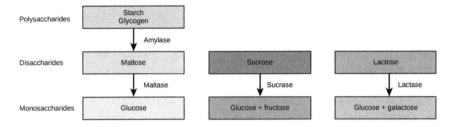

FIGURE 5.7 Digestion of starch and glycogen. Digestion of carbohydrates is performed by several enzymes. Starch and glycogen are broken down into glucose by amylase and maltase. Sucrose (table sugar) and lactose (milk sugar) are broken down by sucrase and lactase, respectively. (Open access article distributed under the Creative Commons Attribution License (4.0 International).)[23]

4. In the large intestine (colon): Carbohydrates that are not digested and absorbed by the small intestine reach the colon where they are partly broken down by intestinal bacteria.

Enzyme activity is partially preserved until reaching the stomach.[2] However, most starch is digested in the small intestine by enzymes from the pancreas. Degradation products of amylose are maltose and maltotriose, while amylopectin degradation produces dextrins and oligomers formed by α-1,6 linkages. Until the end of the intestine is reached, all of these oligomers are degraded to glucose by enzymes such as α-glycosidase and oligo-α-1,6-glucosidase.

In the last century, starch present in foods was considered to be completely digested.[24] However, during the 1980s, studies on starch digestion started to show that besides digestible starch, which could be rapidly or slowly hydrolyzed, there was a variable fraction that resisted hydrolysis by digestive enzymes. That fraction was named resistant starch (RS) and it encompasses those forms of starch that are not accessible to human digestive enzymes but can be fermented by the colonic microbiota, producing short chain fatty acids. Short chain fatty acids from the RS have different physiological and probiotic effects. The scientific interest in RS has increased during the last decades because of its capacity to produce high levels of butyrate throughout the colon. The RS associated with small chains of fructooligosaccharides acts synergistically in the digestive system to cause a *prebiotic* effect that benefits human health. RS has been classified into five types, depending on the mechanism governing its resistance to enzymatic hydrolysis as a result of the transformation process.

The first group (RS I) is the product of treatments in which starch is physically inaccessible and the breakdown of the granular structure does not occur.[2] The second group (RS II) consists of gelatinized starch (i.e., the starch has lost its crystallinity and is composed primarily of amylose); this type is very common in most starchy foods. RS III is formed during starch retrogradation, which occurs after manufacturing in the presence of water, upon cooling, and during storage. Chemical modifications to produce gelling and emulsification agents result in RS IV. Starch containing amylose–lipid complexes and requiring high temperatures of gelatinization forms RS V, which is insoluble in water.

Owing to the interest of the food industry, methods to increase the RS content of isolated starches were developed. Nowadays, the influence of RS on the gut microbiota is a relevant research topic owing to its potential health-related benefits.[24]

5.4 UNCONVENTIONAL STARCHES

The overall starch market is continuously expanding and the current demand is covered by four conventional sources: wheat, corn, potato, and cassava.[2] There are significant differences in the properties of starch of these conventional groups. Starch is also extracted to a lesser extent from rice, barley, sorghum, and sago.[25]

Starches present in legumes, rhizomes, herbs, and seeds are considered unconventional and may be used as ingredients in the same manner as starches from cereals and tubers because of their similar physicochemical and functional properties.[2] These properties are improved by modification treatments and may be employed to develop new processes and new products.

Developments in the starch industry have increased interest in the identification of new starches with distinct properties and their potential for processing at a large scale.[26] Unconventional starches show similarities or advantages with respect to some conventional sources, such as corn, potato, or tapioca, making them attractive additives for food formulations in which the retrogradation process is undesirable; in addition, their aqueous suspension gelatinization requires less time and energy.

5.5 MODIFIED STARCHES

5.5.1 INTRODUCTION

Most starches in their native form have limitations that make them less than ideal for the diversity of desired applications.[27] For this reason, most of the starch utilized as a food or industrial ingredient is first modified, without changing the granular structure, to alter and improve the physical properties of starch polymers in accordance with the intended end use. Modified starches, also called starch derivatives, are prepared by physically, enzymatically, or chemically treating native starch to change their properties.[28,29] Starch can be modified to enhance its positive properties and eliminate deficiencies in its native characteristics. Modified starches are used in practically all starch applications, such as in food products as a thickening agent, stabilizer, or emulsifier; in pharmaceuticals; or as binder in coated paper.

Modification processes can improve the characteristics of native starch by altering its physicochemical and structural properties and increasing its technological value.[2] The characteristics of starch depend on the modification used and are necessary for its industrial use; they include cold water solubility, viscosity and swelling capacity after cooking, retrogradation tendency, and loss of structural order after gelatinization. The industry of starch modification is constantly evolving. Starch is a highly flexible polymer and there are several ways to modify its structure and obtain a functional product with adequate properties for specific industrial applications, increasing its benefit.

Physical modification of starch can increase its water solubility and reduce the size of the starch granules.[2] Physical methods for the treatment of native granules include combinations of temperature and moisture, pressure, shear, and irradiation. Physical modification is simple, inexpensive, and safe. Physical modifications are preferred because they do not require chemical or biological agents that may be harmful.

Chemical modification of starch involves the introduction of functional groups to the molecule without affecting the morphology or granule size distribution.[2] Cationization modifies the dielectric properties of granules depending on their substitution degree, reducing the paste temperature and increasing its viscosity. Crosslinking of polymeric chains increases the degree of polymerization in starch granules, modifying its solubility in organic solvents and reducing its swelling capacity. Acetylation results in the esterification of starch, increasing its swelling capacity and solubility. Acid hydrolysis and oxidation reduce the degree of polymerization and paste viscosity.

Not all starch is digestible, and the indigestible portion is part of the fraction of dietetic fiber or RS.[2] Chemical modifications, such as crosslinking, are used to increase the amount of RS, and these starches are included in paper and textile processes.

The starch industry is in constant expansion, and modification processes increase its versatility.[2] When starch is physically or chemically modified, it can be adapted for different purposes in food and/or non-food industries. Applications of starch modifications can increase the use of unconventional starches and vegetal residues containing starch in the industry. Depending on cost and accessibility, the use of conventional starch can be replaced by that of unconventional starches in industrial processes. Determining the required characteristics of starch for each process is necessary to select the best modification method.

5.5.2 Physical Modification of Starch

Physical modifications of starches are starch property modifications imparted by physical treatments that do not result in any chemical modification of the starch

TABLE 5.1
Non-Thermal Modifications of Starch[3,4]

Technique	Brief Description
Pulsed electric field	Process that uses short pulses of electricity for microbial inactivation
Corona electrical discharge	Electrical discharge brought on by the ionization of a fluid surrounding a conductor
Vacuum ball mill	Type of grinder used to grind and blend materials to make tiny particles
Milling	Micronization
Instant controlled pressure drop (DIC)	Treatment based in a linked intervention of a high temperature–short time treatment and an instant autovaporization by abruptly dropping the pressure toward vacuum[32]
Freezing and thawing	The treatment performed once increases the crystallinity of starch granules, but multiple freezing and thawing cycles could cause irreversible disruption of the crystallinity[33]
Freeze-drying	Low-temperature dehydration process, which involves freezing the product, lowering pressure, then removing the ice by sublimation[34]
High-pressure treatment	High-pressure treatment applied to starch
Thermally inhibited treatment (dry heating)	Heat-treated in a dry state[35]
Superheated starch	Heating and cooling starch suspensions toward spreadable particle gels[36]
Iterated syneresis	Repeated extraction of a liquid from a gel[37]
Sonication	Application of sound waves

other than limited glycosidic bond cleavages.[30] They can be classified into thermal and non-thermal processes. The thermal processes include pre-gelatinization and hydrothermal processes.[31] Non-thermal processes include the use of high hydrostatic pressure, ultrasound, pulsed electric field, and microwave treatment (Table 5.1).[3,4]

5.5.2.1 Thermal Processes

5.5.2.1.1 Pre-gelatinized Starch

Pre-gelatinized starches are starches that have undergone gelatinization and consequently are depolymerized and fragmented, and the granular structure is entirely destroyed because of cooking.[4] The pre-gelatinization process is achieved by different techniques such as drum drying, spray drying, and extrusion cooking. The properties associated with pre-gelatinized starch permit instant dissolution in cold water without heating. Because of the harsh treatment (gelatinization and severe drying) used to obtain pre-gelatinized starch, it is porous and possesses higher water absorption index and water solubility index than those of native starch. Pre-gelatinized starch has been reported to be amorphous and the irregular starch granules of the native starch altered to concave spherical shape by pre-gelatinization. The higher indices of pre-gelatinized starch were explained by their higher macromolecular disorganization, degradation, and weaker associative forces.

Pre-gelatinized starch with various degrees of gelatinization and degradation could be obtained through extrusion. According to Zhu,[38] drum-dried starch possessed much-increased water absorption, swelling, and solubilization than other methods of pre-gelatinized starch production.

There are certain limitations associated with pre-gelatinized starch, which have reduced its applications in certain foods.[4] These include grainy texture and inconsistent and weak gels. These weaknesses have been overcome by the development of granular cold water swelling starch. The latter can exhibit cold water thickening despite keeping its granular integrity, possesses higher viscosity, more homogeneous texture with higher clarity, and has more processing tolerance than pre-gelatinized starch. Contrary to native starch, pre-gelatinized starch and granular cold water swelling starch can rapidly absorb water and increase their viscosity at ambient temperature. This useful functionality has made them applicable in a range of products synthesized at low temperature containing heat-labile components (e.g., vitamins and coloring agents) and instant food. Pre-gelatinized starch and granular cold water swelling starch are also used in cold desserts, instant baby food, pie fillings, gravies, soups, and sauces.

The functional and physicochemical properties of various modified starches determine their applications in the food industry. Pre-gelatinized starches have been used in thermally sensitive foods and as thickeners in many instantaneous products, such as baby food, instant soups, and desserts because of their ability to immediately form pastes when dissolved in cold water.

5.5.2.1.2 Hydrothermal Modifications

Annealing and heat-moisture treatment are the two hydrothermal treatments that modify the physicochemical properties of starch without destroying the granular structure.[4] The similarity of the two processes is that they take place at a temperature above the T_g and below the gelatinization temperature, i.e., the melting temperature (T_m).[4,39] So the structural molecular integrity of the starch granules is preserved in both cases because they are operating at a temperature that is below the disorder temperature. They differ in the water content and temperature. Annealing is the treatment of starch in excess (>60% w/w) or at intermediate (40%–55% w/w) water content, whereas starch treatment below 30%–35% (w/w) is called heat moisture treatment. The hydrothermal modification has significant effect on starch functionality. The following physicochemical properties are affected by both processes: granule morphology and crystallinity, double helix content, amount of amylose–lipid complexes, gelatinization, pasting, swelling power and solubility, gel properties, and susceptibility to acid and enzymatic hydrolysis.

1) **Annealing**. Annealing is a physical process that has been used for centuries to modify the desirable properties of materials, from metals and plastics to biopolymers such as nucleic acids and starches.[40] Annealing results in obvious changes in gelatinization characteristics, such as the size, amylose content, and swelling power of starches, and makes the starch crystalline structures perfect.

The annealing treatment improves molecular mobility.[2] It is associated with a physical reorganization of the starch granule in the presence of water, which is a suitable plasticizer for starch. These changes generate movements in the crystalline and amorphous regions, and physicochemical modifications increase chain interaction in the crystalline region.

Annealing has important industrial applications, imparting different characteristics to products due to an increase in starch granule size, thermal stability, gelatinization temperature, and the availability of starch to digestion by enzymes such as α-amylase.[2]

2) **Heat-moisture treatment.** The physical properties of heat-moisture treated starches depend on the starch botanical source and the treatment conditions used.[4] The treatment brings alterations in functional properties such as a decrease in starch swelling power, solubility, amylose–lipid complexes, and peak viscosity but an increase in the pasting temperature of heat-moisture treated starches.

Heat-moisture treatment can affect the structure and physicochemical properties of cereal, tuber, and legume starches. Generally, heat-moisture treated starches tend to bring about higher gelatinization temperature, lower paste viscosity, a decrease in granular swelling, and an increase in thermal stability. The most important reported effect of heat-moisture treatment is the shift in crystalline structure from B- to A-type for potato and yam starches and a transition from C-type to A-type for sweet potato starch. However, some starches such as corn starch and rice starch were resistant to changes in crystallinity because of heat-moisture treatment.

5.5.2.2 Non-Thermal Physical Processes

Some of the merits of the non-thermal physical treatments of starches include decreased energy utilization, less consumption of water, and extended shelf life of processed food, thereby enhancing global food security.[4] These technologies cause biological, chemical, and physical modifications leading to alterations in sensory, textural, and nutritional properties. Furthermore, these non-thermal technologies retain freshness, nutritional value, and sensory characteristics of food items without any significant thermal degradation.

Non-thermal physical modification is an alternative to the traditional heating processes.[4] The high-energy traditional heat treatments usually diminish cooking flavors and causes loss of vitamins and essential nutrients in the desired product. Compared to traditional thermal processes, the non-thermal processes kill most pathogenic or spoilage microorganisms and inactive enzymes, but minimize the loss of color, taste, texture, nutrients, and heat-labile functional components of food.

Some of the non-thermal processes are conducted at high hydrostatic pressure, use ultrasound effect, pulsed electric field treatment, and microwave treatment.

1) High hydrostatic pressure is a non-thermal food processing technology that decreases undesirable chemical reactions, which may result in undesired organoleptic properties and impose nutritionally adverse effects.[4] High pressure involves using a uniform pressure throughout a product. In the food industry, pressure ranging from 400 to 900 MPa can be used. High hydrostatic pressure treatment inhibits the swelling power of starch granules, so that their viscosity is lower than heat-processed starches. Additionally, starch gelatinization is obtainable at ambient temperature or below 0°C with such a treatment.

2) Ultrasound food processing technology uses frequency in the range of 20 kHz to 10 MHz. Ultrasound is produced with either piezoelectric or magnetostrictive transducers that generate high-energy vibrations. These vibrations are amplified and transferred to a sonotrode or probe, which is in contact with the fluid to be treated. Some merits of ultrasound utilization in food processing are time reduction, energy efficiency, and eco-friendliness. Other advantages of ultrasound are reduction of processing temperature, batch or continuous process, increased heat transfer, deactivation of enzymes, and possible modification of food structure and texture. High-power ultrasound is very significant in the following fields of food processing: filtration, crystallization, homogenization, extrusion, de-foaming, viscosity alteration, separation, emulsification, and extraction. These operations are very important in the separation of gross product into its various components.

3) The pulsed electric field is a non-thermal food preservation method, which kills pathogens or spoilage microorganisms and inactive enzymes, minimizes the loss of taste, color, texture, and nutrients, and heats labile functional components of food.[4] Other advantages of the technology are no toxicity, short treatment time, and that it kills vegetative cells. Various treatments affect the physicochemical properties of starches differently. The applications of pulsed electric field in the food industry result in food spoilage reduction, enhance food safety by increased shell life, and retain freshness of food commodities.

4) Modification of starch by microwave treatment depends on several interacting factors, such as irradiation, furnace dimensions, and the characteristic of the starch.[2] In the microwave irradiation process, the most important parameters are moisture and temperature, which influence the dielectric properties of the starch. Starch modification by microwaves results from the rearrangement of starch molecules that generates changes in solubility, swelling capacity, rheological behavior, T_g, and enthalpies. Depending on the starch source and moisture, modification by microwave also produces variations in morphology and crystallinity in the granule.

5.5.3   ENZYMATIC MODIFICATION OF STARCH

Common enzymes used in starch processing include α-amylase, β-amylase, glucoamylase, pullulanase, and isoamylase.[41] During the hydrolysis process, enzymes attack α-(1→4) and/or α-(1→6) linkages, depolymerizing starch into glucose, maltose, and/or oligosaccharides. Starch is capable of forming inclusion complexes with hydrophobic molecules; however, native starch displays limited capability because of its tendency to retrograde and the highly branched structure of amylopectin. Efforts to enhance starch complexing ability have been carried out through chemical and enzymatic modifications. Enzymatic modification of starch has been used to increase the linear starch content, thus increasing its complexing ability.

The oldest use of enzymes in the pulp and paper industry is in the modification of starches for surface sizing and coating.[42] This practice dates back to the early 1970s and is now carried out more frequently within the paper mills. Starch imparts many beneficial properties to paper, including strength, stiffness, and erasability.

Various enzymatic modifications of starch have been attempted for the applications to the food industry.[43] The major targets of molecular modification of starch by enzymes include the amylose content, the molecular mass, and the structure of amylopectin chains. The main approaches are the indirect *in vitro* method using the carbohydrate-hydrolyzing enzymes from microorganisms or the direct *in vivo* method suppressing or over-expressing the enzymes in the transgenic plants. Microbial enzymes have the potential to modify starch and starch-based foods by hydrolysis, debranching, and/or disproportionation reactions. *Maltogenic* amylase from various bacteria has shown that the enzyme could hydrolyze amylose readily, but hardly attack amylopectin. Thus, maltogenic amylase has great potential in producing starch with different amylose contents. Chain distribution of amylopectin can be engineered by *4-α-glucanotransferase* that disproportionates the side chains of glucan, which can alter the side chain length. The molecular size of starch can also be controlled by the selective hydrolysis reaction with 4-α-glucanotransferase that preferentially cleaves the α-1,4-glycosidic linkage of the glucan segment between the amylopectin clusters. As a result, the apparent mass can be reduced to the level of an amylopectin cluster. Microbial debranching enzymes play a role in shaping glycogen in the cells. Debranching enzymes can be used to modify amylopectin. The enzymatic modification of starch molecules affects properties of the modified starch especially in freeze–thaw stability of gels and retardation of retrogradation during storage.

5.5.4 CHEMICAL MODIFICATION OF STARCH

Food processors generally prefer starches with better behavioral characteristics because of the limitations of native starch characteristics, such as low shear resistance, thermal resistance, and thermal decomposition during processing.[44] Chemical modification of starch, generally achieved through derivatization, such as etherification, esterification, and crosslinking, is employed to optimize the structural characteristics and functional and nutritional properties for targeted applications. The addition of chemically modified starch with low or slow glycemic features might help lower the overall glycemic load of the foods, such as breads and cakes, where significant amounts of readily absorbed forms of carbohydrates are present. Some new applications of chemically modified starches include their use as a chelator, cryoprotectant, drying aid, fat replacer, flavor carrier, and flavor and color precursor.

There exist large numbers of hydroxyl groups on starch molecules, providing the active sites for chemical modification.[45] The source of starch and conditions and methods used for its modification have a great effect on the properties of the modified starch (Table 5.2).

5.5.4.1 Crosslinking

Crosslinking is one of the widely used chemical modifications.[46] Different crosslinking agents are available. When starch is crosslinked, its properties vary. The extent of crosslinking can be predicted with the help of properties like viscosity and swelling. Crosslinked starch, prepared through the reaction of hydroxyl on the starch molecule and the compound with two or more functional groups, shows a three-dimensional network structure.[18] Crosslinking reagents include formaldehyde, epichlorohydrin (EPI), sodium tripolyphosphate (STPP), sodium trimetaphosphate (STMP), adipate, mixed acetic adipic acid anhydride, and phosphorous oxychloride ($POCl_3$) (Figure 5.8).

Crosslinking improves some properties of native starch such as thermo-mechanical shearing, paste stability in acidic medium, gelatinization temperature, and viscosity of starch.

TABLE 5.2
Chemical Modifications of Starch and Their Properties[18]

Modification	Properties
Crosslinking	Better freeze–thaw stability and granule stability; lower swelling power, solubility, and enthalpy of gelatinization; higher heat and shear stabilities
Grafting	Better biodegradability and thermal stability; higher hydrodynamic radius and hydrodynamic volume; stronger water absorbency
Esterification	More moisture resistance and thermoplasticity; lower thermal stability and biodegradation; better compatibility
Etherification	Better thermal stability and solubility; better flowability, permeability, and strength
Oxidization	Higher whiteness, solubility, viscosity, and gelatinization temperature; lower enthalpy values; better water absorbency
Acidification	Better recovery and solubility; lower pasting viscosity; decreased gelatinization enthalpy

Carmona-Garcia et al.[47] investigated the influence of the crosslinked reagent type on morphological, physicochemical, and functional characteristics of banana starch. They found that in crosslinking using EPI and STMP, both reagents attacked the inner part of the starch granules, and while using $POCl_3$ crosslinking only occurred on the surface of the granule. The different morphologies of the native and cross-linked starch granules are shown in Figure 5.9.

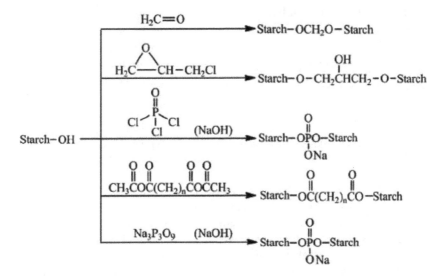

FIGURE 5.8 Some common crosslinking reactions of starch.[46]

FIGURE 5.9 SEM images of starch granules of native starch (a), crosslinked starch with $POCl_3$ (b), crosslinked starch with sodium trimetaphosphate/sodium tripolyphosphate (c), and crosslinked starch with epichlorohydrin (d). (Reproduced with permission from the Royal Society of Chemistry.)[46]

In comparison with native starch, the surface of crosslinked starch became slightly rough and there appeared some black zones (see arrows). The presence of these black zones indicate slight fracture and the formation of deep gutters on the surface of starch granules. They also confirmed that crosslinking decreased protein, moisture, and fat contents of the native starch, while increasing the ash level. The crosslinked starch has the possibility of mass production.

5.5.4.2   Starch Graft Copolymers

Many research groups have investigated graft copolymers based on starch.[18] Synthetic or natural polymers have been introduced into the starch backbone. The change in the structure of the native starch results in some new properties. The main sites for initiation of graft copolymerization are reported to be localized at C1–C2 end groups and C2–C3 glycol groups.

5.5.4.2.1   Grafting with Vinyl Monomers

Starch can be grafted with vinyl monomers such as styrene, acrylic acid, and acrylamide.[18] The grafting reaction mechanisms include free radical, condensation–addition, and ionic interaction mechanisms. Among these mechanisms, free radical is the most commonly used. It means that in the presence of an initiator, starch can generate free radicals, and it then reacts with vinyl monomers to produce free radical sites. Finally, graft copolymers can be achieved by chain growth.

Cerium salts have been used as initiators in graft polymerization (Figure 5.10).

Persulfate is also a widely used initiator, such as potassium persulfate and ammonium persulfate. Fe^{2+}/H_2O_2 was also used as an initiator system to prepare the graft copolymerization of starch with acrylic acid. Starch graft copolymers have great market prospects as flocculants. Additionally, some methods like "click" onto grafting and radiation grafting have been used in grafting copolymerization.

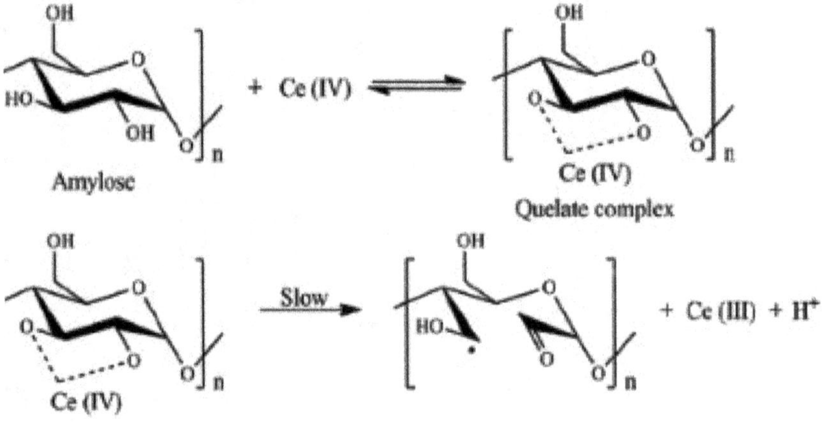

FIGURE 5.10   The radical mechanism process of starch by using Ce(IV) ion.[18]

5.5.4.2.2 Grafting with Other Monomers

Apart from vinyl monomers, there are many other monomers such as *p*-dioxanone, ε-caprolactone, and L-lactide. Starch grafted with these aliphatic polyesters gives excellent properties to the copolymer such as biocompatibility, biodegradability, and amphipathy. Furthermore, some researchers studied starch-grafted graphene.

5.5.4.3 Esterification

Starch can be esterified by carboxylic acid derivatives or acids because of the presence of a large number of hydroxyl groups in the molecule.[18] For the synthesis of starch esters with a high degree of substitution, reactions should be carried out in organic solvents. Generally, there are two types of starch esters: inorganic and organic.

5.5.4.3.1 Inorganic Starch Esters

The starch phosphate is one kind of the inorganic ester and one kind of anionic starch. In comparison with the native starch, starch phosphate possesses higher viscosity, glueyness, and transparency, and therefore it can be used as an adhesive, thickener, stabilizer, and drug-bulking agent. Starch sulfate is another important starch inorganic ester, which got attention due to its biological activities such as anti-tumor, anti-virus, anti-cruor, and other functions.

5.5.4.3.2 Organic Starch Esters

Various kinds of organic starch esters have been synthesized via esterification of starch with anhydride (such as octenyl succinic anhydride, acetic anhydride, propionic anhydride) and chloride.[18]

Some new methods have been developed for starch esterification, including an enzyme method carried out under milder conditions with no negative effects on the environment.

Ionic liquids, as green solvents, get considerable attention owing to their specific properties. They are salts in the liquid state with melting points below 100°C. Researchers studied the enzymatic synthesis of starch laurate in ionic liquids.

5.5.4.4 Etherification

Starch ethers can be prepared by the reaction of starch hydroxyl groups with reactive substances to obtain cationic starch, anionic starch, amphoteric starch, and non-ionic starch.[18] In comparison with native starch, etherified starch shows some useful physio-chemical properties like thermal stability, ion activity, higher reaction efficiency, and thixotropy. Starch ethers can be applied in many areas such as coating, flocculants, drug delivery, additives, papermaking, and so on.

Cationic starch is an important modified starch, in which the positive ionic charge has been produced by the introduction of sulfonium, ammonium, amino, imino, or phosphonium groups. Cationic monomers like 2,3-epoxypropyltrimethyl ammonium chloride or 3-chloro-2-hydroxypropyltrimethyl ammonium chloride are usually used in conventional preparative methods of cationic starch under extrusion, semi-dry, or wet process.

Anionic starch can be prepared by introducing anionic substituent groups like carboxymethyl and sulfonic groups in starch, which shows better properties such as solubility in cold water, strong hydrophilicity, and flocculation. Carboxymethyl starch is the main product among anionic starch and has attracted a lot of attention in both research and industries. CMS can be prepared by various methods like the dry method, aqueous method, and organic solvent method.

Amphoteric starch with the same biodegradability and degradability as native starch means that both cationic and anionic groups are introduced onto the starch molecules. Usually, typical anionic groups in amphoteric starch are phosphonate, phosphate, sulfonate, sulfate, and carboxyl groups, whereas the cationic ones are quaternary ammonium and tertiary amino groups.

Alkyl ether starch is the main non-ionic starch including hydroxypropyl starch and hydroxyethyl starch (HES). HES has received more attention owing to its blood replacement therapy in surgery and intensive care medicine. Nowadays, HESs with high-molecular weight and DS (degree of substitution) have been replaced by HESs with low-molecular weight and DS, which can maintain a potent expansion effect, reduce accumulation of plasma and tissue, and decrease the damage to coagulation system and kidney function.

5.5.4.5   Dual Modification

Classical methods for chemical modification of starch are crosslinking, grafting, esterification, and etherification.[18] However, sometimes single modification cannot meet the requirements. So the development of dual modifications is needed, including a combination of chemical and physical or chemical and enzymatic methods. Dual modifications like acetylation/oxidation, crosslinking/esterification, or crosslinking/oxidation have been mainly investigated. They have been widely used in the food industry as emulsifiers as well as in the non-food industry as adsorbents.

5.5.4.6   Other Chemical Modifications

5.5.4.6.1   Oxidation

Oxidation is a useful modification method to produce oxidized starch and enhances the applications of starch.[18] The most commonly used oxidants include potassium permanganate, sodium hypochlorite, hydrogen peroxide, persulfate, and so on. It is reported that the oxidation reaction primarily occurs at the hydroxyl groups of C-2, C-3, and C-6 positions on a glucosyl unit, *oxidizing the starch hydroxyl groups to carbonyl groups, and then to carboxyl groups. In comparison with the original starch, oxidized starch shows lower viscosity, better stability and film-forming, and lower molecular size. Dialdehyde starch is an important oxidized starch. Ozone as a green* oxidant can replace classical chemical oxidation.

5.5.4.6.2   Acid Modification

Acid modification is another modification method, using acid to modify native starch below the gelatinization temperature. After acid modification, molecular weight, swelling power, solubility, pasting properties, and water binding capacity of all starch decrease. Acid-modified starch has applications in food, paper, textile, pharmaceutical, and other industries.

5.5.5 Thermoplastic Starch

Starch in its granular state is generally unsuitable for thermoplastic processing.[48,49] To obtain thermoplastic starch (TPS), the semi-crystalline starch granules have to be broken down by thermal and mechanical forces. As the melting point of pure starch is higher than its decomposition temperature, plasticizers (water, glycerol, glycol, sorbitol, and so on.) have to be added. The natural crystallinity can then be disrupted by mixing (shearing) at elevated temperatures, which yields TPS. However, this blend is still unsuitable as a bioplastic owing to its high moisture sensitivity and poor mechanical properties. To overcome these drawbacks, starch is chemically modified and/or blended with other (bio)polymers to produce a tougher and more ductile and resilient bioplastic.

Starch-based bioplastics are mainly used for food packagings such as cups, bowls, bottles, cutlery, egg cartons, and straws.[49] Other applications include disposable bags and trash liners as well as compostable films for agriculture. Three main methods are used for TPS processing: casting, extrusion, and thermocompression.[49]

Novamont, an Italian company, is a leader in the world of bioplastics and especially in the area of TPS-based products. Novamont's Mater-Bi is a family of fully biodegradable and compostable bioplastics. Thermoplastic Mater-Bi is processed by the most common conversion technologies: blowing, casting, extrusion/thermoforming, and injection moulding.[50] Raw materials for its production include starches, cellulose, and vegetable oils.

Mater-Bi's biodegradability and compostability properties and its high content of renewable raw materials allow optimal organic waste management and reduce the environmental impact with advantages throughout the entire production–consumption–disposal cycle.

Novamont promotes a bioeconomy model based on the efficient use of resources and territorial integration.[51] Starting from the local areas, they set up biorefineries integrated with the territories and dedicated to the production of bioplastics and bioproducts of renewable origin, conceived to protect soil and water, by revitalizing industrial sites, respecting local specificities and working together with all the players of the whole value chain.[52]

5.6 CONCLUSIONS AND PERSPECTIVES

5.6.1 TARGETED PROPERTIES OF STARCHES

Starches are more and more manufactured and developed in a sustainable and environmental way. Targeted functional properties of starches depend on the application sector where they are used:

- Food and feed sectors: good jellification, stability, spreadability, crispiness, smoothness, and the appropriate viscosity.
- Pharmaceutical sector: tableting, versatility, binding, flowability, and disintegrating powers.
- Paper and other industrial sectors: fluidity, solubility, ionicity, adhesivity, strength, gloss, whiteness, biodegradability, flocculation, and absorption.[53]

5.6.2 Improvement of the Properties of Thermoplastic Starch

Besides the advantages (processability, flexibility, biodegradability, and so on.) of TPS, it also has disadvantages like poor mechanical properties, water sensitivity, poor dimensional stability, and so on.[54] To improve the mechanical properties of TPS-based materials, other additives can also be applied including emulsifiers, cellulose, plant fibers, bark, kaolin, pectin, and others.

One way to improve the properties of starch is to modify them by adding layered silicates and forming so TPS nanocomposites. Polymer/clay nanocomposites are assumed to exhibit improved barrier, thermal, and mechanical properties compared to traditional composites.

Another way to improve the properties of TPS is to reinforce it with natural fibers. All kinds of fibers have been used as reinforcements in TPS including various forms of cellulose and wood. Stiffness and strength increase both with increasing fiber length and with content.

The preparation of polymer blends is a further approach to improve the properties of TPS. The biocompatibility and biodegradability of TPS can be retained if it is blended with biopolymers. One of the biopolymers used and studied the most frequently in such blends is PLA.

5.6.3 Food Starch Market

The **global food starch market** between 2019 and 2024 is poised to register a compound annual growth rate of 5.85%.[55] The market is competitive and driven by an increase in the trend of "health and wellness" and growing consumer demand for all natural ingredients. Cereals, particularly breads and breakfast cereals, and potatoes make the major contribution to starch intakes in the western diets. Starch intakes are considerably higher in countries dependent on starchy staples to a greater degree, such as China and Japan.

Increasing advancements in the technology of microencapsulation have widened the options for the **starch industry**, which is projected to have a high impact in the coming years.[28] The rising demand for weight management ingredients or foods and functionally diverse ingredients is driving the market. These factors are highly impactful all through and are expected to drive the market at a medium pace during the forecast period.

The increasing prices of agricultural commodities highly affect the prices of the final product, which is a major drawback for the market growth.[28] The demand for starch from other industries, such as cosmetics, animal feed, and chemicals, is lagging the **food starch market growth**. However, the rising demand for processed and convenience foods has opened the gates for the market. In addition, the emerging trend of "clean label ingredients" has also triggered the market and is expected to continue the impact throughout the period 2019–2024.

Native and modified starches are used as thickeners, binders, and stabilizers in the food and beverage industry.[28] Owing to their functional properties, they are considered as indispensable ingredients in the industry. Some of their key functions include the prevention of undesired hydration, rendering the desired texture and mouthfeel, shelf-life extension, and encapsulation of other ingredients. The use of modified starch as a fat replacer is expected to further increase its usage in the future because of the growing trend of consumers giving preference for healthy foods, which contain substitutes for fat to increase the nutritive value for food.

The **food starch market** has been segmented by product (native, modified, starch derivatives, starch sweeteners), source (corn, wheat, potato, others), application (beverages, baked goods, confectionery, dairy products, processed foods), and region (North America, South America, Europe, the Middle East and Africa, and Asia-Pacific).[28]

Globally, corn has been the most preferred raw material for starch production.[28] In the United States, more than 90% of the starch is manufactured from corn. Asia-Pacific also follows suit and depends significantly on corn for the manufacture of starch. North America is expected to have the highest market share for the **modified starch market** globally in 2017. The United States is expected to have the highest share of about 45%. Corn is used considerably in Europe as well, although the dependence is lesser compared to other global markets. This is mainly because wheat and potato are significantly used as raw material sources to meet the starch requirements in Europe. The European starch industry produces over 600 products, from native starches to physically or chemically modified starches, through to liquid and solid sweeteners. United Kingdom is one of the prominent countries that have a high market share in the consumption of starch in the region.

The major players in the global food starch market are Agrana, Cargill Incorporated, Grain Processing Corporation, National Starch Food Innovation, Roquette Frères, Tate & Lyle, and Tereos Syral.[28] New entrants in the food starch market are rare, owing to high initial costs and market saturation. Companies are participating in this market with the focus on establishing themselves as manufacturers of health and wellness ingredients. The major competitive factors in the native and modified starches market are price, product customization, and customer service.

REFERENCES

1. B. Pfister and S.C. Zeeman, *Cell Mol. Life Sci.*, 73, 2781, 2016 in https://www.ncbi.nlm.nih.gov/pmc/articles/PMC4919380/
2. S.C. Alcazar-Alay, M.A.A. Meireles, *Food Sci. Technol.*, 35, 215, 2015 in http://www.scielo.br/pdf/cta/v35n2/0101-2061-cta-35-2-215.pdf
3. H.O. Egharevba, *Chemical Properties of Starch*, IntechOpen, Open access book, 2019, doi: 10.5772/intechopen.87777 in https://www.intechopen.com/online-first/chemical-properties-of-starch-and-its-application-in-the-food-industry
4. A. Sarkar and S. Perez, *A Database of Polysaccharides 3D Structures*, CERMAV, 2012 in http://polysac3db.cermav.cnrs.fr/discover_starch.html
5. Birefringence in https://en.wikipedia.org/wiki/Birefringence
6. D.B. Murphy, K.R. Spring, T.J. Fellers and M.W. Davidson, Microscopy U, *Principles of Birefringence* in https://www.microscopyu.com/techniques/polarized-light/principles-of-birefringence
7. Shodhganga, Inflibnet, V.P. & R.P.T.P. Science College in https://shodhganga.inflibnet.ac.in/bitstream/10603/170400/9/09_chapter2.pdf
8. J. Bemiller and R. Whistler Eds., *Starch Chemistry and Technology*, 3rd edition, Academic Press, 2009 in https://www.sciencedirect.com/book/9780127462752/starch

9. S. Perez, P.M. Baldwin and D.J. Gallant, *Structural Features of Starch Granules I in Starch: Chemistry and Technology*, 3rd edition, Academic Press, 2009 in https://www.researchgate.net/profile/Serge_Perez/publication/221658057_Structural_Features_of_Starch_Granules_I/links/5c93b76a299bf111693e2201/Structural-Features-of-Starch-Granules-I.pdf

10. C.G. Biliaderis, *Structural Transitions and Related Physical Properties of Starch in Starch: Chemistry and Technology*, 3rd edition, Academic Press, 2009 in https://www.sciencedirect.com/science/article/pii/B9780127462752000082 and in https://kundoc.com/pdf-structural-transitions-and-related-physical-properties-of-starch-.html

11. A. Sarkar and S. Perez, *A Database of Polysaccharide Structures* in http://polysac3db.cermav.cnrs.fr/discover_starch.html

12. A. Mizuno, M. Mitsuiki and M. Motoki, *J. Agric. Food Chem.*, 46, 98, 1998 in https://pubs.acs.org/doi/10.1021/jf970612b

13. P. Liu, L. Yu, X. Wang, D. Li, L. Chen and X. Li, *J. Cereal Sci.*, 51, 388, 2010 in https://www.sciencedirect.com/science/article/abs/pii/s073352101000041x

14. R. Partanen, L. Murtomaki, T. Moisio, M. Lahteenmaki, O. Toikannen, R. Hartikainen and P. Forssell, *Jpn. J. Food Eng.*, 15, 61, 2014 in https://www.jstage.jst.go.jp/article/jsfe/15/2/15_61/_pdf

15. R.G.M. Van Der Sman and M.B.J. Meinders, *Soft Matter*, 7, 429, 2011, doi: 10.1039/c0sm00280a in https://research.wur.nl/en/publications/prediction-of-the-state-diagram-of-starch-water-mixtures-using-th and in https://www.researchgate.net/publication/255750671_prediction_of_the_state_diagram_of_starch_water_mixtures_using_the_flory-huggins_free_volume_theory

16. Cermav, Lessons, *Starch* in http://Applis.Cermav.Cnrs.Fr/Lessons/Starch/Page.Php.13.Html

17. A. Lovegrove, C.H. Edwards, I. De Noni, H. Patel, S.N. El, T. Grassby, C. Zielke, M. Ulmius, L. Nilsson, P.J. Butterworth, P.R. Elis and P.R. Shewry, *Crit. Rev. Food Sci. Nutr.*, 57, 237, 2017 In https://www.tandfonline.com/doi/full/10.1080/10408398.2014.939263

18. H. Goesaert, K. Brijs, W.S. Veraverbeke, C.M. Courtin, K. Gebruers and J.A. Delcour, *Trends Food Sci. Technol.* 16, 12, 2005 in https://www.sciencedirect.com/science/article/abs/pii/s0924224404001906?via%3dihub

19. Elsevier, ScienceDirect, *Starch Gelatinization* in https://www.sciencedirect.com/topics/food-science/starch-gelatinization

20. *Starch Gelatinization* in https://en.wikipedia.org/wiki/starch_gelatinization

21. S. Kaiser, *Healthy Eating*, 2018 in https://healthyeating.sfgate.com/steps-digestion-carbohydrates-4053.html

22. GKToday, *Digestive System*, 2016 in https://www.gktoday.in/gk/digestive-system/

23. C Molnar, *Concepts of Biology, Digestive System Processes*, 2019 in https://opentextbc.ca/biology/chapter/15-3-digestive-system-processes/

24. L.A. Bello-Perez, P.C. Flores-Silva, E. Agama-Acevedo and J. Tovar, *J. Sci. Food Agric.*, 2018, doi: 10.1002/Jsfa.8955 in https://www.ncbi.nlm.nih.gov/pubmed/29427318

25. Alfa-Laval, *The All-Round Choice for Starch Equipment* in https://www.alfalaval.com/globalassets/documents/industries/food-dairy-and-beverage/starch-and-sweetener/solutions-for-processing-starch.pdf

26. A.L. Santana and M. Angela A. Meireles, *Food Public Health*, 4, 229, 2014 in http://article.sapub.org/pdf/10.5923.j.fph.20140405.04.pdf

27. A. Ashok, M. Mathew and Rejeesh C.R., *IJPSE*, 2, 20, 2016 in https://www.researchgate.net/publication/303312307_innovative_value_chain_development_of_modified_starch_for_a_sustainable_environment_a_review

28. https://en.wikipedia.org/wiki/modified_starch

29. S.C. Alcazar-Alay, M.A.A. Meireles, *Food Sci. Technol.*, 2015 in http://www.scielo.br/pdf/cta/v35n2/0101-2061-cta-35-2-215.pdf

30. J.N. Bemiller, Chapter 5: Physical Modification of Starch, *Starch in Food*, 2nd edition), *Structure, Function and Applications*, Woodhead Publishing Series in Food Science, Technology and Nutrition, 223, 2018 in doi: 10.1016/b978-0-08-100868-3.00005-6

31. A.O. Ashogbon, *Global Nutr. Diet.*, 1, 001, 2018 in https://scientonline.org/open-access/current-research-addressing-physical-modification-of-starch-from-various-botanical-sources.pdf

32. N. Albitar, S. Mounifr and K. Allaf, Association Française De Séchage Pour L'industrie Et L'agriculture - Afsia, Lyon, May 2009 in https://www.researchgate.net/publication/281880674_The_Instant_controlled_pressure_drop_DIC_technology_as_a_manufacturing_process_of_high_Quality_dried_onion

33. M.A. Villar et al (Eds), *Starch-Based Materials in Food Packaging*, Academic Press, Elsevier, 2017

34. https://en.wikipedia.org/wiki/freeze-drying

35. S.T. Lim, H.S. Lim, J. Han, J.N. Bemiller, *Cereal Chem.*, 79, 601, 2002 in https://www.researchgate.net/publication/237328303_Modification_of_Starch_by_Dry_Heating_with_Ionic_Gums

36. P.A.M. Steeneken and A.J.J. Woortman, *Food Hydrocolloids*, 23, 394, 2009 in https://www.sciencedirect.com/science/article/pii/s0268005x08000155

37. G. Lewandowicz and M.S. Smietana, *Carbohydr. Polym.*, 56, 403, 2004 in http://citeseerx.ist.psu.edu/viewdoc/download?doi=10.1.1.452.1251&rep=rep1&type=pdf

38. F. Zhu, *Carbohydr. Polym.*, 122, 456, 2015, doi: 10.1016/j.carbpol.2014.10.063 in https://www.ncbi.nlm.nih.gov/pubmed/25817690

39. H.F. Zobel, S.N. Young and L.A. Rocca, *Cereal Chem.*, 65, 443, 1988 in https://pdfs.semanticscholar.org/6bd2/a47f3da45e6dbb8000532216a15c5d292bcf.pdf

40. K. Yu, Y. Wang, Y. Wang, L. Guo and X. Du, *Int. J. Food Prop.*, 19, 1272, 2016 in https://www.tandfonline.com/doi/pdf/10.1080/10942912.2015.1071842

41. A.I. Gonzalez Conde, Theses and Dissertations, *Effects of Chemical and Enzymatic Modifications on Starch and Naringenin Complexation*, University of Arkansas, Fayetteville, 2017 in https://pdfs.semanticscholar.org/081e/b752ecde13cead9b7c8b3a611596c7f46449.pdf

42. P. Bajpai, Biotechnology for Pulp and Paper Processing, Enzymatic Modification of Starch for Surface Sizing, 431, Springer, 2018 in https://link.springer.com/chapter/10.1007/978-981-10-7853-8_19

43. K.H. Park, J.H. Park, S. Lee, S.H. Yoo and J.W. Kim, Enzymatic Modification of Starch For Food Industry, *Carbohydrate-Active Enzymes, Structure, Function and Applications*, Woodhead Publishing Series in Food Science, Technology and Nutrition, 157, 2008 in https://doi.org/10.1533/9781845695750.2.157

44. Y.F. Chen, L. Kaur and J. Singh, Chapter 7: Chemical Modification of Starch, *Starch in Food*, 2nd edition, Woodhead Publishing Series in *Food Science, Technology and Nutrition*, 283, 2018 in https://doi.org/10.1016/b978-0-08-100868-3.00007-x

45. Q. Chen, H. Yu, L. Wang and Z.U. Abdin, *RSC Adv.*, 5(83), doi: 10.1039/C5ra10849g in https://www.researchgate.net/publication/280973931_recent_progress_in_chemical_modification_of_starch_and_its_applications

46. N. Shah, R.K. Mewada and T. Mehta, *Rev. Chem. Eng.*, 32, 265, 2016 in https://www.researchgate.net/profile/Nimish_Shah4/publication/299498738_Crosslinking_of_starch_and_its_effect_on_viscosity_behaviour/links/5aa1fa5ba6fdcc22e2d26c02/Crosslinking-of-starch-and-its-effect-on-viscosity-behaviour.pdf

47. R. Carmona-Garcia, M.M. Sanchez-Rivera, G. Mendez- Montealvo, B. Garza-Montaya and L.A. Bello-Perez, *Carbohydr. Polym.*,, 76, 117, 2009 in https://books.google.be/books?id=-L0YAwAAQBAJ&pg=PT413&lpg=PT413&dq=R.+CARM

ONA-GARCIA,+M.M.+SANCHEZ-RIVERA,+G.+MENDEZ-+MONTEALVO,
+B.+GARZA-MONTAYA+and+L.A.+BELLO-PEREZ,+Carbohydr.+Polym.,+76,+1
17,+2009&source=bl&ots=u0a1Vk9vZ2&sig=ACfU3U092TVMPyyEqw8AjxRCY7
R5qxT9eQ&hl=fr&sa=X&ved=2ahUKEwjcxPjaoqToAhXQzKQKHadhDVIQ6AEw
AHoECAkQAQ%23v=onepage&q=R.%20CARMONA-GARCIA%2C%20M.M.%20
SANCHEZ-RIVERA%2C%20G.%20MENDEZ-%20MONTEALVO%2C%20B.%2-
0GARZA-MONTAYA%20and%20L.A.%20BELLO-PEREZ%2C%20Carbohydr.%20
Polym.%2C%2076%2C%20117%2C%202009&f=false#v=snippet&q=R.%20
CARMONA-GARCIA%2C%20M.M.%20SANCHEZ-RIVERA%2C%20G.%20
MENDEZ-%20MONTEALVO%2C%20B.%20GARZA-MONTAYA%20and%20
L.A.%20BELLO-PEREZ%2C%20Carbohydr.%20Polym.%2C%2076%2C%20
117%2C%202009&f=false

48. *Polymerdatabase*, 2018 in https://polymerdatabase.com/polymer%20brands/starch.
html

49. A. David, PhD Thesis, *Etude De Dérivés De L'amidon: Relation Entre La Structure Et
Le Comportement Thermomécanique*, Lille University, 2017 In https://ori-nuxeo.univ-
lille1.fr/nuxeo/site/esupversions/a9136500-fef7-4097-b5f0-2e2cfac6d25c

50. Novamont in https://www.novamont.com/eng/mater-bi

51. Novamont in https://www.novamont.com/eng/

52. Novamont in https://uk.novamont.com/

53. Roquette in https://www.roquette.com/product-overview/starches/

54. P. Muller, PhD Thesis, *Structure-Property Correlations in Thermoplastic Starch Blends
and Composites*, Budapest University of Technology and Economics, 2015 in https://
repozitorium.omikk.bme.hu/bitstream/handle/10890/1492/ertekezes.
pdf?sequence=2&isallowed=y

55. *Mordor Intelligence*, 2018 in https://www.mordorintelligence.com/industry-reports/
food-starch-market

6 Applications of Starch in the Bioeconomy

6.1 INTRODUCTION

Starch-containing crops form an important constituent of the human diet and a large proportion of the food consumed by the world's population originates from them.[1] Besides the use of the starch-containing plant parts directly as a food source, starch is harvested and used as such or chemically or enzymatically processed into a variety of different products such as starch hydrolysates, glucose syrups, fructose, starch or maltodextrin derivatives, or cyclodextrins. Despite a large number of plants that are able to produce starch, only a few plants are important for industrial starch processing. The major industrial sources are maize, tapioca, potato, and wheat.

6.2 STARCH INDUSTRY

6.2.1 INDUSTRIAL STARCH MARKET: OVERVIEW

Starch is used in a variety of industrial applications.[2] Food industry accounts for a significant share in terms of consumption in the global industrial starch market. Industrial starches are used in manufacturing various products in the food industry such as bakery products, confectioneries, canned jams and fruits, commercial caramel, and monosodium glutamate. Industrial starch is also used in the non-food industries such as paper, textile, mining, building materials, and consumer product industries. It is used in the paper industry during the manufacturing and coating process. In the textile industry, it is used for cloth printing and finishing. In the mining industry, it is used in the well-drilling process. Starch is mixed with clay and this mixture provides proper water holding ability and viscosity to drill oil wells.

Based on the product type, the global industrial starch market can be segmented as cationic starch, ethylated starch, oxidized starch, acid-modified starch, and unmodified starch.[2] Based on applications, the global industrial starch market can be segmented as corrugating, building materials, papermaking, paper coating, mining/drilling, multiwall bag, liquid detergents, and biofuel and biomaterial.

The global industrial starch market can be divided into seven regions, namely North America, Latin America, Western Europe, Eastern Europe, Asia Pacific Excluding Japan (APEJ), Japan, and the Middle East and Africa.[2] APEJ holds a major share in the industrial starch market in terms of consumption. This is attributed to the growth of end-use industries such as textile, paper, and food industries in the emerging clusters of the region.

6.2.2 Starch Production and Main Outputs in the World

The principal industrial productions of starch are based only on four main resources, maize, cassava, wheat, and potatoes, which represent 76%, 12%, 7%, and 4% respectively.[3] The other botanical resources represent less than 1%. The main production areas are North America, China, Europe, Southeast Asia, and South America. The market is led by North America, China, and Europe, which represent in total 85% of the world production. This geographical predominance is also represented by the top 10 companies (Table 6.1).

The main starch outputs are for food and non-food applications.[3] Food products represent 60% of the market with 31% and 29% for confectionery and drinks, and processed food, respectively.[5] The feed is around 1%. Non-food products represent 29% with 19%, 6%, and 4% for corrugating and papermaking, pharmaceuticals and chemicals, and other non-food products.

About 35% of all starch in Western countries are used in its native or modified form as biopolymers in the food, textile, and paper manufacturing industries (Figure 6.1).[6] The use of maize starch in the United States exemplifies this; most of

TABLE 6.1
Top 10 Starch Companies in the World[3,4]

Cargill	USA
Ingredion	USA
ADM	USA
Tate & Lyle	UK
Roquette	France
Zhucheng Xingmao	China
Global Bio Chem	China
Tereos	France
COFCO	China
Xiwang	China

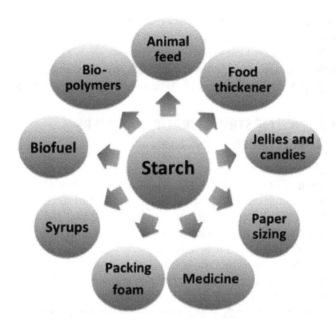

FIGURE 6.1 Main uses of converted starch.[6] Native starches may be modified chemically, physically, enzymatically, or by a combination of these processes to create a material that is specifically designed to one of these uses. (Reproduced with permission from Taylor & Francis.)

it is not directly consumed but converted into a diverse set of products, ranging from animal feeds to sweeteners, to polymers, and fuels.

Many starch-based biopolymers and starch-derived biofuels are projected to be less harmful to the environment than those derived from petrochemicals.

6.2.3 Starch Production and Main Outputs in Europe

The European starch industry produces over six hundred products, from native starches to physically or chemically modified starches, through to liquid and solid sweeteners.[7] The versatility of starch products is such that they are used as ingredients and functional supplements in a vast array of food, non-food, and feed applications. From 77 starch production facilities in 20 of the 28 European Union (EU) member states, the European Starch Industry today produces 10.8 million tons of starch each year from EU wheat, maize, and potatoes. EU consumption of starch and starch derivatives was 9.4 million tons in 2017.

Key figures for the EU starch market are[4]:

- The companies, members of the trade association Starch Europe, process starch in 77 plants in 20 of the 28 EU member states.
- They process about 24 million tons of agricultural raw materials (17 million tons of cereals and 7 million tons of starch potatoes) into 11 million tons of starch and 5 million tons of co-products.
- EU starch production has increased from 8.7 million tons in 2004 to 10.8 million tons in 2017.

- The EU consumes 9.4 million tons of starch (excluding starch by-products totaling around 5 million tons), of which 58% are in food, 2% are in feed, and 40% are in non-food applications, primarily paper making.
- Of the 9.4 million tons of starch and starch derivatives consumed in the EU, 27% are native starches, 19% are modified starches, and 54% are starch sweeteners.

6.3 INFLUENCE OF STRUCTURAL PARAMETERS OF STARCH ON ITS FUNCTION

Starches of different molecular structures are exploited to create products with a broad continuum of functionalities.[6] The proportion of amylose to amylopectin, the amount of lipid complexed with amylose, the ratio of AtoB glucan chains in amylopectin, glucan chain helical conformation, and granule size and morphology collectively dictate the properties of starch. They do so by altering starch gelatinization, viscosity, swelling power, and retrogradation.

The amylose-to-amylopectin ratio is often a key target for starch improvement.[5] The amylose content is negatively correlated with granule sizes. As a result of having little to no amylose, waxy starch is more susceptible to digestion than normal and high-amylose starch (Table 6.2). It also requires less time and lower temperatures to become gelatinized, producing a clear paste, which remains viscous over time. Waxy starch is, therefore, ideal for use as a food thickener and as an adhesive in many industries. In contrast, high-amylose starches, which are resistant to digestion, may serve as a source of dietary fiber, providing low-glycemic-index carbohydrates for individuals who require strict glycemic management.

These examples illustrate how genetic alterations in starch enzymes can profoundly influence the end-use of starch.[6] This spectrum of uses of starch is further extended by modifying native starches enzymatically or chemically, and by subjecting it to physical treatments.

There are two broad methodologies for manipulating plants through biotechnological means: *reverse genetics*, which are targeted approaches, and *forward genetics*,

TABLE 6.2

Applications of Waxy Starch, Normal Starch, and High-Amylose Starch in Food and Non-Food Industry[5]

	Waxy Starch (Little to Low Amylose)	Normal Starch	High-Amylose Starch
Food industry	Thickener	Beverage	Edible film
	Freeze–thaw stabilizer	Brewery	Frying batter
	Emulsifier	Confectionery Bakery	Sausage casing Confectionery
Non-food industry	Paper	Paper	Bioplastic
	Textile	Pharmaceutical	Corrugated cardboard
	Adhesive	Cosmetic	Resistant starch/dietary fiber
	Livestock feed	Textile	Colon drug delivery
		Biofuel	Prebiotic

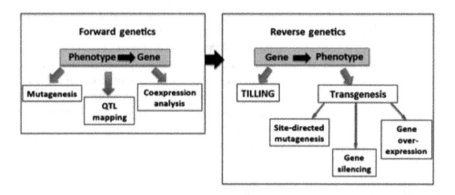

FIGURE 6.2 Biotechnological approaches to altering the genes that may control starch biosynthesis.[6] These may be broadly defined as either the forward or the reverse genetics approach. TILLING, targeting-induced local lesions in genomes; QTL, quantitative trait loci. (Reproduced with permission from Taylor & Francis.)

which are open-end approaches (Figure 6.2). The forward genetics approach uses variation in the trait among different individuals and then seeks to identify the causal gene. The reverse genetics approach starts with a candidate gene of unknown function but based on indicative sequence motifs or domains present, it may be predicted to be involved in a particular process(es), which is then tested. Often, candidate genes identified by forward genetics are verified using reverse genetics.

Gene sequences that may encode starch metabolic enzymes can be identified using bioinformatics and their functions tested by reverse genetics methods such as transgenic modifications and *targeting-induced local lesions in genomes (TILLING)*.[6] The *transgenic modification* changes the genetic makeup of an individual by introducing deoxyribonucleic acid (DNA) into its genome. TILLING is a high-throughput Polymerase Chain Reaction (PCR)-based method of screening a mutant collection for alterations in the sequence of a gene of interest.

The forward genetics approach toward manipulating starch biosynthesis would start by connecting an observed variation in a starch trait among a group of related organisms to an underlying gene(s) or alleles of that gene.[6] There are several forward genetics approaches, among which are mutagenesis, quantitative trait loci mapping, and coexpression analysis. Once identified, the causal genes or alleles of the genes can be integrated into the commercially important cereal varieties through breeding or TILLING, thus establishing a cereal genotype with desirable starch traits.

6.4 APPLICATIONS FOR THE FOOD INDUSTRY

6.4.1 Overview

It has been decades since the commercial production of starch for food and industrial applications was initiated.[8,9] The biological function of starch is a reserve of carbon and energy.[10] As a food, starch is the most abundant and important digestible polysaccharide. The starches in food are commonly derived from grains or seeds (corn, wheat, rice, and barley), tubers (potato), and roots (cassava). Starch provides 70%–80% of the calories consumed by humans worldwide.

Native starches do not meet the physicochemical and functional features required by foods, so they are modified.[11] Modified starches are used for food applications. Maize, wheat, and rice are used to modify the texture and appearance of foods. Starch modifies the adhesiveness, thickening, glazing, emulsion stability, binding, clouding, foam stability, moisture retention, dusting, expansion, crisping, gelling, and edible films and coatings. Therefore, it is widely used in foods (Table 6.3). Maize starch is widely used because of its pasting properties, which depend on the amylose content.

Waxy maize starch produces a bright and translucent paste with a weak structure, whereas high-amylose starch produces an opaque and stiff paste that is used in the production of hard gels used in gum candies. Normal maize starch is used to produce sweeteners and fat substitutes.

Maize starch is modified to improve functionality; waxy maize starch is hydrolyzed with acid or enzymes to produce nanocrystals that are used to stabilize emulsions; also, it can be modified with n-octenyl succinate acid to improve its hydrophobic character and water-holding capacity. Such starches are used as bread improvers because they can substitute for the fat, providing a soft texture. Waxy maize starch is pregelatinized by physical treatments to use in jelly confectionery.

Wheat starch is used for moisture control, thickening, and adhesiveness in batters, ice creams, soups, gravies, and dressings; also, it is added to yogurt to improve thickening and gelling characteristics; in sausages and comminuted meats, it is used to improve the water-binding properties. Wheat starch is used to produce glucose syrup and sweeteners in the beverage and confectionery industries.

Rice starch has diverse applications because it can be obtained with a wide range of amylose-to-amylopectin ratios, producing pastes with different texture characteristics. Rice starch is a gluten-free ingredient; therefore, it is widely used in this

TABLE 6.3
Functionality and Uses of Cereal Starches[11]

Starch	Functions	Uses
Maize	Thickener, binder, filling, stabilizer, gelling, food additive	Sweetener products, thick sauce, smooth food texture, glutinousness, formed meat, confectionery fillings, candies and batters, jam-filled waffles, emulsifiers, encapsulation of antioxidant products
Wheat	Gas by fermentation and rigid network binder, thickener, adhesive agent	Bakery products, batters, ice cream, soups, gravies, dressings, yogurts
Rice	Binder, fat mimetic, freeze–thaw stability, whiteness, dusting; crispness agent, thickener	Confectionery, pastries, puddings, custards, smooth gravies, sauces, soups, snacks, ice cream, baby foods, freeze–thawed cake
Barley	Gelling and thickness agent	Desserts
Oat	Fat replacer and emulsifier	Frozen desserts, ice creams, instant breakfast drinks, dressings, gravies

kind of products. Rice starch presents excellent mouthfeel, is a substitute for fat, and shows freeze–thawing stability with low syneresis, producing a spongy structure with a slow change in the texture of gels; in this sense, it is used in frozen products such as freeze–thawed cakes.

Barley starch is used as a gelling and thickening agent because after heating at 96°C for several minutes, the viscosity is not modified.

The internal lipids in oat starch produce amylose–lipid complexes that retard its retrogradation and solubility; oat starch is used in cheese making as a fat substitute.

6.4.2 EDIBLE FILMS

In the food industry, edible films are barriers that prevent moisture transfer, gas exchange, oxidation, and the movement of solutes, while maintaining their organoleptic properties.[10]

Edible films can be produced from materials with film-forming ability.[12] During manufacturing, film materials must be dispersed and dissolved in a solvent such as water, alcohol, or a mixture of water and alcohol or a mixture of other solvents. Plasticizers, antimicrobial agents, colors, or flavors can be added in this process. Adjusting the pH and/or heating the solutions may be done for the specific polymer to facilitate dispersion. The film solution is then cast and dried at a desired temperature and relative humidity to obtain free-standing films. In food applications, film solutions could be applied to food by several methods such as dipping, spraying, brushing, and panning followed by drying.

Edible films and coatings have received much attention because of their advantages over synthetic films.[8] The main advantage of edible films over traditional synthetics is that they can be consumed with packaged products. There is no package to dispose of; even if the films are not consumed, they could still contribute to the reduction of environmental pollution. The films produced from renewable materials are anticipated to degrade more readily than conventional synthetic materials. Their main disadvantage lies in their mechanical and permeable properties. Edible films require starches with a high amylose content (≥70%).[10] The amylopectin molecule cannot adequately form films; the branched structure imparts poor mechanical properties to the film, reducing its tensile strength and elongation.

Polysaccharides are typically hygroscopic and therefore are poor barriers to moisture and gas exchange.[10] The use of plasticizers in the film composition improves the barrier against moisture exchange and restricts microbial activity. Native starch does not produce films with adequate mechanical properties and requires pretreatment, the use of a plasticizer, mixing with other materials, genetic or chemical modification, or a combination of these treatments. Among the plasticizers, for hydrophilic polymers, such as starch, are glycerol and other low-molecular-weight polyhydroxy-compounds, polyether, and urea. Processes such as extrusion adjust the parameters of temperature and mechanical energy over the starch paste, making it a thermoplastic material that is also suitable for the production of edible films.

6.5   APPLICATIONS FOR THE NON-FOOD INDUSTRY

The traditional major non-food usage of starch is in paper and paperboard, textiles, and fermented products.[13] Starch acts as an adhesive, thickener, and film-forming, coating, and sizing agent in paper and textile products. The production of ethanol from starch using a fermentation process started most probably in beer-producing countries approximately in the 12th century.[14] Today, bioethanol from starch has many uses. Blended with petrol, it makes transport fuel more sustainable.[15,16] Most of the U.S. production of fuel ethanol is from corn starch.

Recently, significant progress has been made in the development of biodegradable plastics, including starch, to produce biodegradable materials with similar functionality to that of oil-based conventional polymers. These bio-based materials have several benefits for greenhouse gas balances and in the use of renewable, rather than finite resources. The use of biodegradable materials will contribute to sustainability and reduction in the environmental impact.

Starch is also used in novel materials, which have the desired mechanical properties through plasticization and blending, but also touches on the use of chemical modification to achieve the right property balance.[17]

6.5.1   PAPER AND TEXTILES

In the paper industry, starch is used as a flocculant and retention aid, as a bonding agent, as a surface sizing agent, as a binder for coatings, and as an adhesive in corrugated board.[8] In 2010, consumption of industrial corn starch for paper and paperboard production in the U.S. exceeded 1.1 million metric tons, of which 40% are chemically modified starch.

The textile industry is a major user of modified starches.[8] Modified starches are a significant component to develop textile capability (strength, elasticity, and surface strength), reduce friction in the weaving process, and improve the feel and appearance of the fabric.[18] They are specifically used for warp sizing preparatory to weaving, for sizing or finishing the cloth after it is woven, or in printing certain types of fabrics. They are also used in large quantities in laundering, both in the commercial establishment and at home.

6.5.2   BIOETHANOL AND OTHER BIOFUELS

6.5.2.1   Bioethanol

Ethanol is a clear, colorless liquid and the main ingredient in alcoholic beverages like beer, wine, or brandy.[19] Bioethanol is a bio-based ethanol. Because it can readily dissolve in water and other organic compounds, ethanol also is an ingredient in a range of products, from personal care and beauty products to paints and varnishes to fuel. In 2020, bioethanol producers have begun producing hand sanitizers and surface disinfectants from bioethanol at their plants [20]

Production of bioethanol from starch is a process including as major stages: hydrolysis of higher sugars to monosaccharides (e.g., glucose), *fermentation* of glucose to produce ethanol and carbon dioxide, and product separation/purification.[21,22]

Bioethanol is commercially produced in one of the two ways: dry milling and wet milling process.[23] The difference is that wet milling involves separating the grain kernel into its parts, e.g., germ, fiber, protein, and starch, before fermentation, whereas the dry milling process grinds the entire grain kernel into flour, the starch in the flour is converted into ethanol during the fermentation process, emitting carbon dioxide and *dried distillers grains with solubles* (*DDGS*). Dry milling is a simpler process than wet milling, but it also produces fewer products.

From a chemistry point of view, the process can be written as follows, with the first reaction catalyzed by hydrolytic enzymes, such as amylases, and the second catalyzed by enzymes present in yeast or bacteria:

$$(C_6H_{10}O_5)_n + nH_2O \rightarrow nC_6H_{12}O_6 \rightarrow 2nCH_3CH_2OH + 2nCO_2$$

The major steps in the dry milling process are grinding, cooking and liquefaction, saccharification, fermentation, and distillation (Figure 6.3).[24] The main products of dry milling are ethanol, carbon dioxide, and DDGS. Detailed process steps include[23,24]:

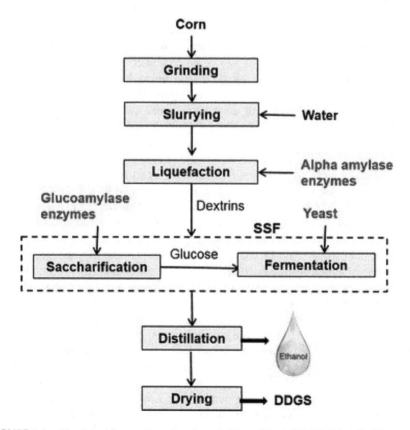

FIGURE 6.3   Bioethanol from processing of starch (dry milling). DDGS, dried distillers grains with solubles; SSF, simultaneous saccharification and fermentation. (Reproduced with permission from Biotechnology for Biofuels (Creative Commons Attribution 4.0 International).)[25]

1. Grinding

   With the dry milling process, the grain is screened to remove debris and ground into a coarse flour. The milled grain is then mixed with water to form a *mash*.

2. Cooking and liquefaction

   Once the mash is made, it goes through cooking and liquefaction. The cooking stage is also called gelatinization. Water interacts with the starch granules in the corn when the temperature is >60°C and forms a viscous suspension. The liquefaction step is actually a partial hydrolysis that lowers the viscosity. It is essentially breaking up the longer starch chains into smaller chains. To accomplish liquefaction, the reaction must take place under certain conditions. The pH of the mash should be maintained in the range of 5.9–6.2, and ammonia and sulfuric acid are added to the tank to maintain the pH. Enzymes such as α-amylase are added in the liquefaction step to convert starch into smaller carbohydrates. Enzymatic hydrolysis may be carried out by using soluble enzymes (the most conventional method) or immobilized enzymes.[19] Recent advances in the development of enzyme immobilization showed promising results.

3. Saccharification

   The next step in the process is saccharification. Saccharification is the process of further hydrolysis to glucose monomers. A different enzyme is used, called a glucoamylase (also known as amyloglucosidase or γ-amylase). It cleaves both the α (1,4) and α (1,6) glycosidic bonds to form glucose. There are a wide variety of amylase enzymes available that are derived from bacteria and fungi. The optimum conditions are different from the previous step and include a pH of 4.5 and a temperature of 55°C–65°C.

4. Fermentation

   The final chemical step is fermentation. Yeast is added to convert the simple sugar to ethanol and carbon dioxide. A common yeast to use is *Saccharomyces cerevisiae*, which is a unicellular fungus. The reaction takes place at 30°C–32°C for 2–3 days in a batch process. Close to 90%–95% of the glucose is converted to ethanol.

   It is possible to do saccharification and fermentation in one step. It is called simultaneous saccharification and fermentation (SSF), and both glucoamylase and yeast are added together. It is done at a lower temperature than saccharification (32°C–35°C), which slows the hydrolysis into glucose. As glucose is formed, it is fermented, which reduces enzyme product inhibition. It lowers initial glucose concentrations, lowers contamination risk, lowers energy requirements, and produces higher yields of ethanol. Because SSF is done in one unit, it can improve capital costs and save residence time.

5. Distillation

   The fermented mash is then pumped into a multicolumn distillation system. The columns use the difference of the boiling point of ethanol and water to separate ethanol from the mash. Ethanol in the stream usually can be in 95% purity with water.

6. Dehydration

 The distilled ethanol is then passed through a molecular sieve to physically separate the remaining water. This is based on the different molecular sizes to produce anhydrous ethanol for storage or usage.

For wet milling, the corn kernels are broken down into starch, fiber, corn germ, and proteins (gluten) by heating in sulfurous acid solution.[22] The starch is separated and can produce ethanol, corn syrup, or food-grade starch. The wet milling process also produces additional products including feed, corn oil, gluten meal (~60% protein), and gluten feed (high fiber content, ~20% protein). Corn gluten feed consists mainly of maize bran and maize steep liquor (liquid separated after steeping). The wet milling process yields maize starch, maize germ oil meal, corn gluten meal, and corn gluten feed as main products.

6.5.2.2 Bio-butanol

Butanol is mainly composed of four isomeric structures, i.e., n-butanol (n-C_4H_9OH), *sec*-butanol (*sec*-C_4H_9OH), *iso*-butanol (*iso*-C_4H_9OH), and *tert*-butanol (*tert*-C_4H_9OH).[26]

 In recent years, n-butanol has caught the attention of researchers as an alternative biofuel to bioethanol. Although most researchers and industries previously focused on ethanol as a fuel rather than butanol, butanol could be a better direct option. The straight-chain alcohol (n-butanol) derived from the biochemical route, the so-called bio-butanol, is sometimes considered as the next generation biofuel because of its many advantages over ethanol, such as higher energy content, lower volatility, and the property of not readily absorbing moisture.

 An attractive route for the production of bio-butanol is through the fermentation of sugar in the presence of different bacteria from the Clostridiaceae family.[24] Acetone–butanol–ethanol (ABE) fermentation has many benefits. It is a process that uses bacterial fermentation to produce acetone, n-butanol, and ethanol from carbohydrates such as starch. The process may be likened to how yeast ferments sugars to produce ethanol but the organisms that carry out the ABE fermentation are strictly anaerobic. The ABE fermentation produces solvents in a ratio of three parts acetone, six parts butanol to one part ethanol.

 Batch butanol fermentation is divided into two stages: sugar conversion into organic acids and solvent production.[27] The bacteria grow exponentially in the first phase of fermentation, producing acids (mostly acetate and butyrate). This leads to a decrease in pH to around 4.5. Toward the end of the first phase, the production rate falls as the bacterial cells shift their activity from acid production to solvent production, in response to the low pH. The acid products induce solvent-producing enzymes necessary for the second stage.

6.5.3 Starch Plastics

Starch plastics, blends of starch with other polymers, are being developed to contribute to environmental problems such as climate change and plastic pollution, owing to their bio-based origin and possible biodegradability.[28,29] They are among the earliest commercialized bio-based plastics and are produced on an industrial scale today. At present, commercial starch plastics are developed mainly for film (e.g., biodegradable packaging, bags, agricultural mulching films), injection molding (e.g., disposable tableware, flowerpots), and foam applications (e.g., loose-fill packaging). Flexible packaging accounts for about half of the starch plastic market, the remainder being used in agriculture, rigid packaging, and consumer goods. The main crops used for native starch production are corn, wheat, potato, and cassava. However, starch is also available in waste streams. For example, in the potato industry, starch-rich waste streams can amount to up to 15 kg for each 100 kg of potatoes processed. Thermoplastic starch (TPS) currently represents the most widely used bioplastic, accounting for a large part of the bioplastics market.[30]

Native starch cannot be used in plastics directly; its granular structure first needs to be disrupted using water, heat, and typically also plasticizers such as glycerol or sorbitol.[31] This process yields TPS, which can be processed like other plastics, e.g., in extrusion or injection molding. However, pure TPS has poor mechanical properties and is susceptible to water, which limits its potential product applications. Therefore, TPS is often compounded with other polymers, typically aliphatic polyesters, to improve mechanical properties. Because hydrophobic polyesters and hydrophilic starch are immiscible, compatibilizer additives also need to be introduced to ensure good adhesion between the components, which improves technical performance. By varying the components used during compounding (native starch, copolymers, and additives), starch plastics with a wide range of technical properties can be obtained. Starch plastics can be biodegradable if TPS is blended with other biodegradable components.

6.5.4 STARCH-BASED COMPOSITES AND NANOCOMPOSITES

6.5.4.1 Starch-Based Polymer Composites

Starch, which is a biodegradable polymer and produced in abundance at low cost, is reported to be one of the most promising candidates for fabrication of bioplastics.[16,32] Because of this, starch has been receiving growing attention since the 1970s.[33] Many efforts have been exerted to develop starch-based polymers for conserving the petrochemical resources, reducing environmental impact, and searching for more applications. Starch-based polymers in particular include those reinforced by lignocellulosic fibers, as biodegradable plastic materials.[34] Numerous studies have been conducted to optimize the properties of starch-based bioplastics. The most important properties in bioplastic materials include mechanical and thermoforming properties, gas and water vapor permeability, transparency, and availability. Final properties of the TPS can be improved by using different fillers, as well as by changing the source of the starch.[35]

Starch-based polymers are being marketed on a commercial scale, by trademarks such as Mater-Bi (Novamont, Italy), Bioplast (Biotech, Germany), Biopar (Biopolymer Technologies AG, Germany), Novon (produced by Chisso in Japan and Warner–Lambert in the USA), Cardia Bioplastics (Cardia Bioplastics Ltd., Australia), and Plantic R1 (Plantic Technologies Ltd., Australia).[35]

6.5.4.2 Starch-Based Polymer Nanocomposites

In recent years, significant attention has been given to a new class of bioplastic materials represented by nanocomposites as a promising alternative to conventional plastics because of their superior properties.[30] Nanocomposites consist of a polymer matrix reinforced with nano-dimensional particles instead of the conventional microdimensional fillers. Researchers at Toyota Company in the 1980s first reported that the presence of nanoparticles in the matrix contributed to significant improvements in both physical and mechanical properties. Later research efforts were devoted to improving the performance of materials through the use of nanoparticles. Clay particles can be effectively used as nanocomposites, owing to their unique structure and properties. When added to form nanocomposites, the clays significantly enhance the mechanical performance of bioplastics, improve their moisture resistance, and reduce the release of plasticizer from starch. This shows how bioplastic nanocomposites would play a role in replacing conventional plastics.

As an interesting case study, the Osaka University has developed marine biodegradable plastics based on starch and cellulose nanofibers.[36] Thanks to a proprietary production process, the finished product is claimed to exhibit excellent water-resistance and high strength, while also being very biodegradable when left floating in seawater over time.

Further developments of starch-based nanocomposites will depend on the theoretical knowledge of the nanostructure materials, the modification of fillers for targeted bioplastics, the mechanisms for superior reinforcement as compared to their micro-counterparts, and the establishment of a simple structure–property relationship. Most nanocomposite fabrication methods have reported nanoparticle agglomeration causing the formation of irregularly shaped nanostructural features within the composite.

REFERENCES

1. S. Shanmugan and T. Sathishkumar, Enzyme Technology, *Enzymes in Starch Industry*, I.K. International Publishing House, 2009
2. Persistence Market Research, *Industrial Starch Market: Global Industry Analysis and forecast 2017–2025* in https://www.persistencemarketresearch.com/market-research/industrial-starch-market.asp
3. P.J. Halley and L. Averous (Eds), *Starch Polymers, From Genetic Engineering to Green Applications*, Starch Production and Main Outputs in the World, Elsevier, 2014 in https://www.elsevier.com/books/starch-polymers/halley/978-0-444-53730-0
4. MarketWatch, 2019 in https://www.marketwatch.com/press-release/global-corn-starch-market-2019-top-manufacturers-regions-market-distribution-industry-size-supply-and-demand-scenario-forecast-to-2025-2019-09-25
5. S.W. Horstmann, K.M. Lynch and E.K. Arendt *Food*, 6, 29, 2017 in https://www.mdpi.com/2304-8158/6/4/29/htm
6. D.M. Beckles and M. Thitisaksakul, Uc Davis, *Use of Biotechnology to Engineer Starch in Cereals*, 2010, doi: 10.1081/e-ebaf-120051354 in https://escholarship.org/uc/item/8pq8x654
7. Starch.Eu, 2019 in https://starch.eu/the-european-starch-industry/

8. A.V. Singh, L.K. Nath and A. Singh, *Electr. J. Environ., Agric. Food Chem.*, 9, 1214, 2010 in https://www.researchgate.net/publication/235694313_Pharmaceutical_food_and_non-food_applications_of_modified_starches_A_critical_review

9. D. Perin and E. Murano, *Nat. Product Commun.*, 12, 837, 2017 in https://www.researchgate.net/publication/317949581_Starch_Polysaccharides_in_the_Human_Diet_Effect_of_the_Different_Source_and_Processing_on_its_Absorption

10. S.C. Alcazar-Alay and M.A.A. Meireles, *Food Sci. Technol.*, Campinas, 35, 215, 2015 in http://dx.doi.org/10.1590/1678-457X.6749

11. M.T.S. Clerici and M. Schmiele (Eds), *Starches for Food Applications*, Elsevier, 2018 in https://www.elsevier.com/books/starches-for-food-application/silva-clerici/978-0-12-809440-2

12. T. Bourtoom, *IFRJ*, 15, 237, 2008 in https://www.researchgate.net/publication/228397700_Edible_films_and_coatings_Characteristics_and_properties

13. B.P. Singh (Ed.), *Industrial Crops and Uses*, CABI, 2010 in https://www.cabi.org/bookshop/book/9781845936167

14. R.A. Meyers (Ed.), *Encyclopedia of Sustainability Science and Technology*, 2012, Springer in https://link.springer.com/referenceworkentry/10.1007%2F978-1-4419-0851-3_432

15. Crop.energies, Südzucker Group, 2017 in http://www.cropenergies.com/Pdf/en/Bioethanol/Verwendung.pdf

16. Decoflame, *Bioethanol* in https://decoflame.com/bioethanol/

17. P.M. Visakh and Y. Long Eds., *Starch-Based Blends, Composites and Nanocomposites*, Royal Society of Chemistry, 2015 in https://pubs.rsc.org/en/content/ebook/978-1-84973-979–5

18. Sonish Starch, *Textile Applications* in http://www.sonishstarch.com/products/product_modified/product_textile/

19. ChemicalSafetyFacts.org, *Ethanol* in https://www.chemicalsafetyfacts.org/ethanol/

20. Argus in https://www.argusmedia.com/en/news/ 2090007-germany-approves-ethanol-use-in-disinfectant-production

21. A.Kang and T.S. Lee, *Bioeng. (Basel)*, 2, 184, 2015 in https://www.ncbi.nlm.nih.gov/pmc/articles/PMC5597089/

22. M. Zabochnicka-Sviatek and L. Slawik, *Archivum Combustionis*, 30, 237, 2010 in https://www.researchgate.net/publication/228351087_Bioethanol-Production_and_Utilization

23. B. Lantz and X. Li, *Bioethanol from Starch Process Simulation Tutorial on PRO II/8.2*, Washington State University, in https://piazza.com/class_profile/get_resource/halehf4tif520p/hd6fwliaouk6vm

24. J.A. Dutton, e-Education Institute, PennState, 2018 in https://www.e-education.psu.edu/egee439/node/673

25. D. Kumar, *Biotechnology for Biofuels*, 9, 228, 2016 in https://www.researchgate.net/publication/309398674_Dry-grind_processing_using_amylase_corn_and_superior_yeast_to_reduce_the_exogenous_enzyme_requirements_in_bioethanol_production

26. B. Ndaba, I. Chiyanzu and S. Marx, *Biotechnol. Rep. (Amst)*, 8(1), 2015 in https://www.ncbi.nlm.nih.gov/pmc/articles/PMC4980751/

27. B. Kolesinska et al., *Materials*, 12, 350, 2019 in https://www.mdpi.com/1996-1944/12/3/350

28. J.H. Song, R.J. Murphy, R. Narayan and G.B.H. Davies, *Biodegradable and Compostable Alternatives to Conventional Plastics* in https://www.ncbi.nlm.nih.gov/pmc/articles/PMC2873018/

29. M.L.M. Broeren, L. Kuling, E. Worrel and L. SHEN, Resources, *Conserv. Recy.*, 127, 246, 2017 in http://www.sciencedirect.com/science/article/pii/S0921344917302793

30. https://en.wikipedia.org/wiki/Bioplastic#Starch-based_plastics

31. N. Jabeen, I. Majid and G.A. Nayik, *Cogent. Food Agric.*, 1(1), 2015 in https://www.tandfonline.com/doi/full/10.1080/23311932.2015.1117749

32. B.R. Mose, and S.M. Maranga, *J. Mat. Sci. Eng. B*, 1, 239, 2011 in https://www.academia.edu/22671661/A_Review_on_Starch_Based_Nanocomposites_for_Bioplastic_Materials
33. D.R. Lu, C.M. Xiao and S.J. Xu, *Express Polym. Lett.*, 3, 366, 2009 in http://www.expresspolymlett.com/articles/EPL-0000946_article.pdf
34. W.N. Gilfillan, Developing Starch-Based Polymer Composites, Thesis, QUT, Queensland, Australia, 2015 in https://eprints.qut.edu.au/86612/6/William_Gilfillan_Thesis.pdf
35. T.N. Prabhu and K. Prashantha, *Polym. Compos.*, 39(7), 2499, 2018 in https://www.researchgate.net/publication/309364948_A_review_on_present_status_and_future_challenges_of_starch_based_polymer_films_and_their_composites_in_food_packaging_applications
36. *Barrett, Osaka University Develops Marine Biodegradable Plastics, Bioplastics News,* 2020 in https://bioplasticsnews.com/2020/03/24/osaka-university-marine-biodegradable-plastics/

7 Perspectives of Starch in the Bioeconomy

7.1 INTRODUCTION

The economy of tomorrow will be more bio-based than today. Simultaneously resolving climate change, oil supply, energy security, and food supply is a key challenge for the humanity. Only the use of new technologies will allow us to bridge the gap between economic growth and environmental sustainability progressively. The bioeconomy uses not only renewable resources such as starch but also municipal solid waste and algae instead of fossil resources.

The need for a circular economy is becoming widely acknowledged[1] (Figure 7.1).

A circular economy aims at maintaining the value of products, materials, and resources for as long as possible by returning them into the product cycle at the end of their use, while minimizing the generation of waste.[3] The fewer the products we discard and the lesser the materials we extract, the better the environment. This process starts at the very beginning of a product's lifecycle: smart product design and production processes can help save resources, avoid inefficient waste management, and create new business opportunities.

Bio-based products are wholly or partly derived from materials of biological origin, excluding materials embedded in geological formations and/or fossilized.[4] In industrial processes, enzymes are used in the production of chemical building blocks, detergents, pulp and paper, textiles, and so on. By using fermentation and biocatalysis instead of traditional chemical syntheses, a higher process efficiency can be obtained, resulting in a decrease in energy and water consumption and a reduction of toxic waste. As they are derived from renewable raw materials such as

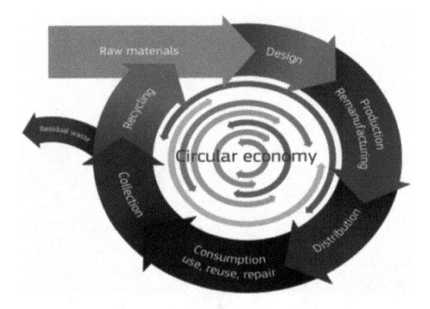

FIGURE 7.1 Circular economy. (Reproduced by courtesy of Centre de contact Europe Direct.)[2]

plants, bio-based products can help reduce carbon dioxide and offer other advantages such as lower toxicity or novel product characteristics.

In October 2018, the European Commission put forward an action plan to develop a sustainable and circular bioeconomy that serves Europe's society, environment, and economy.[5] In November 2018, Starch Europe hosted its annual conference on the theme of innovating together for a sustainable food system.[6] Ongoing discussions around, inter alia, Food2030, CAP Reform, the European Union (EU) Bioeconomy Strategy, and the EU Protein Plan stress the importance of a multi-stakeholder approach to achieving a climate-smart, environmentally sound, and sustainable food system. Starch Europe seeks to facilitate the debate on how continuous innovation helps achieve these goals and the contributions of the starch industry in particular.[7]

This chapter will focus on (1) the sustainability of starch products; (2) thermoplastic starch-based materials; (3) starch for ethanol; (4) starch in ionic liquids; and (5) starch as a feedstock for advanced functional materials.[8,9]

7.2 SUSTAINABILITY OF STARCH PRODUCTS

7.2.1 Vision

The most well-known vision of sustainable development has been the one promoted within the United Nations, from the 1987 Brundtland Commission report to the 1992 Rio Conference and the 2002 Johannesburg Summit on Sustainable Development.[10] In the Brundtland report, sustainable development is defined as the development that meets the needs of current generations without compromising the ability of future generations to meet their own needs.[11]

The sustainability of starch products has been summarized in 2015 by the European starch industry as follows[12]:

- **Economic pillar**:
  - Processing 23 million tons of EU raw materials for production
  - Producing 15.5 million tons of starches, starch derivatives, and co-products
  - Euro 8.3 billion revenues
  - Euro 380 million invested, of which over Euro 140 million are in R&D
- **Environmental pillar**:
  - Raw materials must comply with the EU Common Agricultural Policy's cross-compliance and greening measures
  - Less than 1% waste
  - Wide use of combined heat and power
  - Helps prevent food waste
  - Replaces fossil-fuel ingredients in industrial applications
  - Pioneer in life cycle assessment with sector-wide studies in 2001, 2012, and 2015
- **Social pillar**:
  - 14,600 direct jobs and 100,000 indirect jobs mainly in rural areas
  - Full compliance with all national, EU, and international labor standards
  - EU starch industry safety program to recognize best performers.

7.2.2 EUROPEAN STARCH INDUSTRY

In 2018, the raw materials processed by the European starch industry are split almost equally among wheat, maize, and starch potatoes transformed into 10.7 million tons of starch (Figure 7.2). This represents ~6% of the EU wheat production, ~12% of the EU maize crop, and the totality of the EU starch potato crop. The main applications of starch in Europe are illustrated in Figure 7.3.

The starch production process is a key factor of the environmental pillar. The concept of biorefinery is at the core of the European starch industry's economic development: valorization of biomass, energy and water efficiency, and waste minimization, all contribute to the competitiveness of the European starch industry. The starch industry does not only produce starch; the over-riding objective of starch producers is to valorize all the components of the agricultural raw materials. It processes every part of the plant and produces minimal waste; less than 1% is not valorized.

7.3 STARCH FOR THE PRODUCTION OF RENEWABLE, BIODEGRADABLE BIOPLASTICS

The rapid accumulation of plastic waste is driving an international demand for renewable plastics with superior qualities, such as full biodegradability, as part of an expanding circular bioeconomy (Figure 7.4).[15] Starch, which is a totally biodegradable

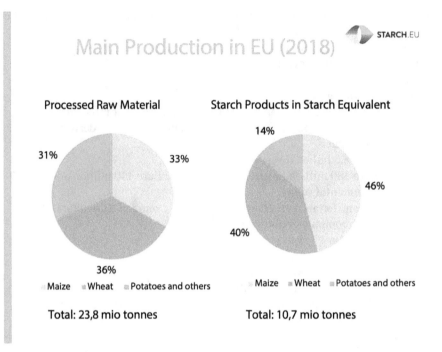

FIGURE 7.2   The European starch industry's production (2018 data). (Reproduced by courtesy of Starch Europe.)[7,13]

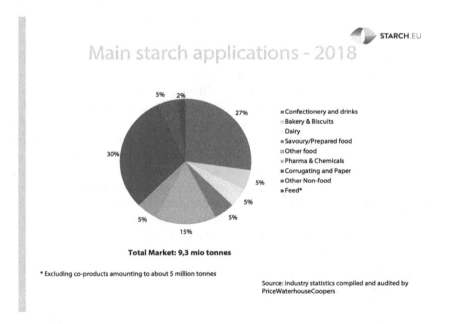

FIGURE 7.3   Main starch applications in Europe (2018 data). (Reproduced by courtesy of Starch Europe.)[7]

FIGURE 7.4   Plastic pollution: discarded plastic bags are a hazard to marine life.[14]

polymer in a wide variety of environments, can be processed into thermoplastic materials and can replace synthetic plastics.[16,17]

The glass transition temperature (Tg) of pure, dry starch is above its decomposition point, and therefore, it does not soften and flow.[11] However, starch can be plasticized by relatively low levels (15–30 wt %) of molecules that are capable of hydrogen bonding with the starch hydroxyl groups, such as water, glycerol, and sorbitol. This thermoplastic starch (TPS) will flow at elevated temperatures and pressures and can be extruded to give both foams and solid molded articles. TPS finds applications in various industries including packaging.[18] Unfortunately, the properties of TPS tend to be disappointing. For example, TPS plasticized with water has poor dimensional stability and becomes brittle if the water content is lost, and the properties of water- and glycerol-plasticized TPS are poor at high humidity.

The properties of TPS can be improved significantly by blending with other polymers, fillers, and fibers.[11] Both natural and synthetic polymers have been used for this purpose, including cellulose, zein (a protein from corn), natural rubber, polyvinyl alcohol, acrylate copolymers, polyethylene and ethylene copolymers, polyesters, and polyurethanes. Blending is usually accomplished by twin-screw extrusion at an elevated temperature. Blends of TPS with other biodegradable polymers, such as polyvinyl alcohol or aliphatic polyesters like polylactic acid, polycaprolactone, and poly(3-hydroxybutyrate), are fully biodegradable. For TPS blends with non-biodegradable polymers, it is likely that only the TPS component will biodegrade in a meaningful timeframe. Reinforced, 100% renewable TPS blends can be obtained by including natural fibers, such as wood pulp, hemp, and other plant fibers.

Further studies that focus on reducing water absorption and decreasing retrogradation of the material to avoid decreasing mechanical strength or stiffness during storage are desirable.[19]

7.4   STARCH FOR THE PRODUCTION OF ETHANOL

Based on average prices and product yields in 2018, a typical dry mill ethanol plant was adding nearly $2 of additional value—or 55%—to every bushel of corn

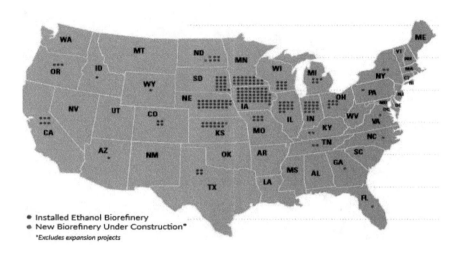

FIGURE 7.5 U.S. fuel ethanol biorefineries by state. (Reproduced with permission from RFA Renewable Fuels Association.)[15]

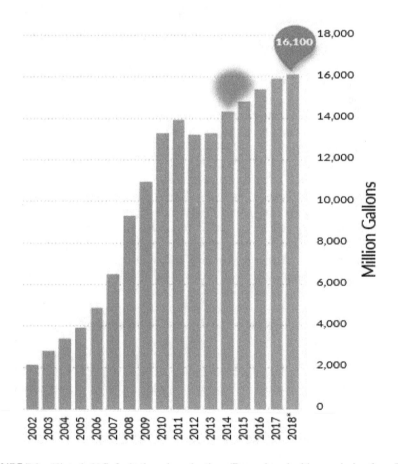

FIGURE 7.6 Historic U.S. fuel ethanol production. (Reproduced with permission from RFA Renewable Fuels Association.)[15]

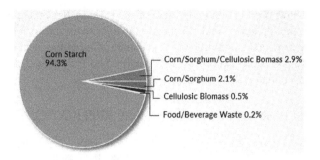

FIGURE 7.7   U.S. ethanol production capacity by feedstock type. (Reproduced with permission from RFA Renewable Fuels Association.)[15]

processed.[29] For a corn cost per bushel of $3.35, the value of outputs per bushel is $3.84 for ethanol, $1.16 for distillers grains, and $0.19 for corn distillers oil.

In 2018, in the United States, 210 ethanol plants located across 27 states (Figure 7.5) produced an astounding 16.1 billion gallons of clean-burning renewable ethanol (Figure 7.6). The total consumption rose to a record 16.2 billion gallons, 300 million gallons more than the previous year.

With new ethanol facilities starting up production and existing facilities expanding, the industry will continue to provide additional employment opportunities and add value to farm products.[14]

The production of ethanol is among man's earliest ventures into value-added processing.[14] Henry Ford and Alexander Graham Bell were among the first to recognize that the plentiful sugars found in plants could be easily and inexpensively converted into clean-burning, renewable fuel. While the production concept remains the same, today's ethanol industry uses state-of-the-art technologies to produce high-octane ethanol and valuable co-products from the starches and sugars. More than 90% of the U.S. fuel ethanol is produced using the dry milling process, with the remaining 9% coming from wet mills. The main difference between the two processes is in the initial treatment of the grain. The ethanol production capacity of the United States is shown in Figure 7.7, with corn starch accounting for 94.3%.

Because of improvements in production efficiencies and the use of "new" feedstocks, today's ethanol biorefinery operates much like a chemical refinery, which is able to produce multiple renewable fuels and products. Some biorefineries are producing biodiesel and renewable diesel from corn distillers oil, but the largest impact has been on corn kernel fiber production.

7.5   STARCH FOR NEW APPLICATIONS OF IONIC LIQUIDS

Ionic liquids (ILs) have been widely recognized as promising "green solvents" to replace volatile organic solvents for polysaccharide processing.[21] Over the past few years, ILs have been increasingly demonstrated to serve as excellent media for the dissolution, plasticization, and derivatization of starch. This allows the synthesis of chemically modified starches with a high degree of substitution and the development of various starch-based materials such as TPS, composite films, solid polymer electrolytes, nanoparticles, and drug carriers.

Ren et al.[21] have presented an overview of the roles of ILs in starch dissolution, gelatinization, modification, and plasticization and their industrial applications. Moreover, they provided a comprehensive understanding of the mechanisms behind the IL-processing of starch and provided insights into the rational development of novel starch-based materials with ILs.

7.6 STARCH FOR NOVEL FUNCTIONAL APPLICATIONS

Several advanced applications of starch have been reported in the literature, for example as self-healing materials, foams, pharmaceutical excipients, and antimicrobial films, and in drug delivery and water treatment (Figure 7.8).[3]

7.6.1 APPLICATIONS OF STARCH IN SELF-HEALING POLYMERIC MATERIALS

Most self-healing polymers and their healants are petroleum based. However, the utilization of bio-based derivatives as healants in polymers has been studied.[22] One interesting study utilized starch as a healant in self-healing polymers. Waxy maize starch (WMS) was used both as a healant and as a polymer. WMS was gelatinized to activate the hydroxyl groups on starch and then encapsulated with poly(D,L-lactide-co-glycolide) using a double emulsion solvent evaporation technique. Starch utilized as a healant in self-healing thermoset resins has been shown to be a viable alternative material to currently used highly reactive and toxic chemicals.

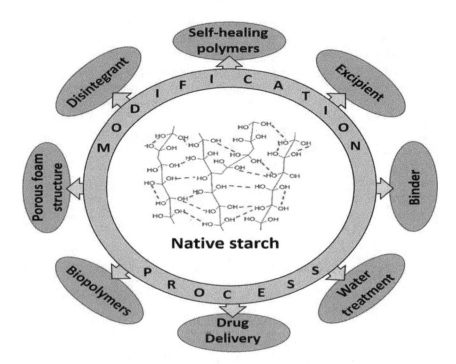

FIGURE 7.8 Advanced material applications of starch and its derivatives. (Reproduced with permission from Elsevier).[3]

7.6.2 Applications of Starch in Porous Foam Structures

The binding ability of starch has given rise to its use as a binder and pore former for metallic and ceramic porous foam structures. Porous starches are important modified starches and are now attracting considerable attention owing to their great adsorption ability.[23] These starches contain abundant pores from the surface to the center of the granules, which increase the specific surface area, and are excellent natural absorbents. There is a growing interest in exploiting their properties in various food and non-food areas.

7.6.3 Applications of Starch in Water Treatment

Recently, natural polysaccharides have been developed as environmentally friendly materials for removing toxic pollutants from aqueous solutions.[24] Among numerous polysaccharides, starch is particularly promising. The main criterion in the design of modified starch with substantial stability is fast complexation of toxic pollutants. A new class of modified starches has been synthesized and applied as adsorbents for dyes and heavy metals.

7.6.4 Pharmaceutical Applications

Starch has attracted substantial interest as a pharmaceutical excipient and as a component in drug delivery.[25]

Owing to its non-toxic nature, non-irritating properties, as well as low cost, ease of modification, and versatility in uses, starch is placed in a leading position among polymers used as pharmaceutical excipients.[26] In many conventional tablets and capsules, starch is used as a diluent, disintegrant, binder, and lubricant.

Starch has also been used for a wide range of specialized drug delivery applications such as delivery of challenging molecules and targeting to specific sites in the body. Although several official native starches with different proprietary identities are available, new sources will continue to evolve with the spate of economic and scientific interests in starch and starch-based products. Starch should continue to be a material of great value in drug delivery because of its overwhelming intrinsic properties, low cost, versatility in application, and ease of modification.

7.6.5 Applications of Starch in Antimicrobial Films and Coatings

The proliferation of antibiotic-resistant bacteria has posed a substantial demand for innovative strategies to fight pathogenic bacterial infection in the health and personal care as well as food contamination.[1] Antimicrobial packaging is one of the most used active packagings, which can reduce or retard the growth of pathogenic and spoilage bacteria in the food matrix.[27,28] Starch-based film is considered a promising material for antimicrobial packaging.[29]

The use of edible films and coatings is an environmentally friendly technology that offers substantial advantages for an increase of shelf-life of many food products.[30] The development of new natural edible films and coatings with the addition of antimicrobial compounds to preserve food products including fruits, vegetables, and seafood is a technological challenge for the industry and is a very active research field worldwide.[31]

REFERENCES

1. S. Leipold and A. Petit-Boix, *J. Clean. Prod.*, 201, 1125, 2018 in https://www.sciencedirect.com/science/article/pii/S0959652618323503

2. European Commission, R2Pi project – *"transition from linear 2 circular: Policy and Innovation"* – in https://ec.europa.eu/easme/en/news/r2-supporting-transition-circular-economy

3 Eropean Commission, Eurostat in https://ec.europa.eu/eurostat/web/circular-economy

4. European Commission, *Internal Market/Industry/Entrepreneurship/SMEs, Sectors, Biotechnology, Bio-based Products* in https://ec.europa.eu/growth/sectors/biotechnology/bio-based-products_en

5. European Commission, *A New Bioeconomy Strategy for a Sustainable Europe*, 2018 in https://ec.europa.eu/commission/news/new-bioeconomy-strategy-sustainable-europe-2018-oct-11-0_en

6. *Starch Europe, Annual Conference*, 2018 in https://starch.eu/starch-europe-conference-2018-2/

7. BlogActiv.eu, 2018 in https://guests.blogactiv.eu/2018/10/30/how-starch-facilitates-the-development-of-a-more-sustainable-agriculture-and-food-system-in-the-eu/

8. E. Ogunsona, E. Ojogbo and T. Mekonnen, *Eur. Polym. J.*, 108, 570, 2018 in https://www.researchgate.net/publication/327814585_Advanced_material_applications_of_starch_and_its_derivatives

9. E.C. Ojogbo, *Starch Modification for Sustainable and Functional Material Applications*, Thesis, Waterloo, Ontario, Canada, 2019 in https://uwspace.uwaterloo.ca/bitstream/handle/10012/14463/Ewomazino_Ojogbo.pdf?sequence=3&isAllowed=y

10. M. Mason, LSE Research Online, The London School of Economics and Political Science, 2011 in http://eprints.lse.ac.uk/36653/1/sustainability_challenge_%28LSE_RO%29.pdf

11. International Institute for Sustainable Development, The Knowledge to Act, *Sustainable Development*, 2019 in https://www.iisd.org/topic/sustainable-development

12. Starch.eu, 2018 in https://starch.eu/the-european-starch-industry/

13. J. Jacques, Starch.eu 2018, *Adding Value with the Bioeconomy* in https://ec.europa.eu/info/sites/info/files/food-farming-fisheries/events/presentations/agri-outlook-day2-session8-jacques_en.pdf

14. Shutterstock. Richard Whitcombe/Shutterstock, CC BY-ND in http://theconversation.com/the-future-of-plastics-reusing-the-bad-and-encouraging-the-good-87001

15. H. Karan, C. Funk, M. Grabert, M. Oey and B. Hankamer, *Trends Plant Sci.*, 2019, doi: 10.1016/j.tplants.2018.11.010 in https://www.researchgate.net/publication/330128246_Green_Bioplastics_as_Part_of_a_Circular_Bioeconomy

16. X.L. Wang, K.K. Yang and Y.Z. Wang, *J. Macromol. Sci., Part C*, 43, 385, 2003 in https://www.tandfonline.com/doi/abs/10.1081/MC-120023911?mobileUi=0&journalCode=lmsc19

17. J Gotro, Polymer Innovation Blog, *Thermoplastic Starch: A Renewable, Biodegradable Bioplastic*, 2013 in https://polymerinnovationblog.com/thermoplastic-starch-a-renewable-biodegradable-bioplastic/

18. T. Niranjana Prabhu and K. Prashantha, *Polym. Compos.*, 2016, doi: 10.1002/pc.24236, https://onlinelibrary.wiley.com/doi/10.1002/pc.24236

19. A.M. Nafchi, M. Moradpour, M. Saeidi and A.K. Alias, *Starch/Stärke*, 65, 61, 2013 in https://www.researchgate.net/publication/258211268_Thermoplastic_starches_Properties_challenges_and_prospects

20. Renewable Fuels Association, Powered with Renewed Energy, 2019 in https://ethanol-rfa.org/wp-content/uploads/2019/02/RFA2019Outlook.pdf
21. F. Ren, J. Wang, F. Xie, K; Zan and S. Wang, *Green Chemistry, Applications of Ionic Liquids in Starch Chemistry: A Review* in https://pubs.rsc.org/en/content/articlelanding/2020/gc/c9gc03738a#!divAbstract
22. .R. Kim, A.N. Netravali, *Polymer*, 117, 150, 2017 in https://www.researchgate.net/publication/316022953_Self-healing_starch-based_%27green%27_thermoset_resin
23. L. Liu, W. Shen, W. Zhang, F. Li, and Z. Zhu, Chap 4, Porous Starch and Its Applications, *Functional Starch and Applications in Food*, Z. Jin (Ed), Springer Nature Singapore Pte Ltd, 2018, in https://www.researchgate.net/publication/327748340_Porous_Starch_and_Its_Applications/citation/download
24. R. Cheng and S. Ou, Polymer Science: Research Advances, Practical Applications and Educational Aspects, p. 52, *Application of Modified Starches in Wastewater Treatment*, 2016 in https://pdfs.semanticscholar.org/0308/b6197e72b211e92b0c5fbc08079b-50f1eb2b.pdf
25. Pharmapproach, Pharmaceutical Technology, 2019 in https://www.pharmapproach.com/pharmaceutical-applications-of-native-starch/
26. P. Builders and M.I. Arhewoh, *Starch/Stärke*, 68(1), 2016 in https://www.researchgate.net/publication/292185896_Pharmaceutical_Applications_of_Native_Starch_in_Conventional_Drug_Delivery
27. ScienceDirect, *Antimicrobial Packaging* in https://www.sciencedirect.com/topics/food-science/antimicrobial-packaging
28. A.E. Oprea and A.M. Grumezescu, *Nanotechnology Applications in Food, Flavor, Stability, Nutrition and Safety*, Academic Press, Elsevier, 2017 in https://www.elsevier.com/books/nanotechnology-applications-in-food/grumezescu/978-0-12-811942-6
29. E. Salleh and I.I. Muhamad, *AIP Conf. Proc.*, doi: 10.1063/1.3377861, 2010 in https://www.researchgate.net/publication/234924001_Starch-based_Antimicrobial_Films_Incorporated_with_Lauric_Acid_and_Chitosan
30. S.A. Valencia-Chamorro, L. Palou, M.A. Del Rio and M.B. Perez-Gago, *Crit. Rev. Food Sci. Nutr.*, 51, 872, 2011 in https://www.ncbi.nlm.nih.gov/pubmed/21888536
31. Food Ingredients First, All News, *Starch-Based, Edible Film for Seafood Could Kill Pathogens on Food Surface*, 2018 in https://www.foodingredientsfirst.com/news/starch-based-edible-film-for-seafood-could-kill-pathogens-on-food-surfaces.html

Glossary/Dictionary

A

Acarbose ($C_{25}H_{43}NO_{18}$): An inhibitor of α-glucosidase, an intestinal enzyme that releases glucose from larger carbohydrates such as starch. It is a trisaccharide.

Acetyl CoA: A water-soluble molecule that carries acetyl groups in cells. It consists of an acetyl group linked to coenzyme A (CoA) by an easily hydrolyzable thioester bond.

ADP (adenosine diphosphate): A nucleotide essential to the flow of energy in living cells.

ADPglucose (adenosine diphosphate glucose): Serves as the glycosyl donor for the formation of bacterial glycogen, amylose in green algae, and amylopectin in higher plants.

α-Amylase: An enzyme that hydrolyzes in a random manner α-bonds of large, α-linked polysaccharides, such as starch and glycogen, yielding glucose and maltose (PDB 1HNY; EC3.2.1.1). Salivary α-amylase, a major component of human saliva, plays a role in the initial digestion of starch.

α-Limit dextrin: Short-chained branched amylopectin remnant, produced by hydrolysis of amylopectin with α-amylase.

Allosteric regulation: Regulation of an enzyme by binding an effector molecule at a site other than the active site of the enzyme.

Amylodextrin: A linear dextrin or short-chained amylose (DP 20–30) that can be produced by enzymatic hydrolysis of the α-1,6-glycosidic bonds or debranching amylopectin.

Amyloplast: A non-pigmented organelle found in some plant cells. It is responsible for the synthesis and storage of starch granules, through the polymerization of glucose. Amyloplast also converts this starch back into sugar when the plant needs energy.

Apposition: Gradual centrifugal deposition of layers.

ATP (adenosine triphosphate): A nucleotide that provides energy to drive many processes in living cells.

Autotroph: An organism that can produce its own food.

B

(β/α)₈ barrel fold: The (β/α)₈ barrel is the most common fold among protein catalysts. The barrel structure is composed of eight alternating strand-loop-helix-turn units. The β-strands are located in the interior of the protein, forming the staves of a barrel, whereas the α-helices pack around the exterior.

β-Amylase: An enzyme that catalyzes the hydrolysis of 1,4-α-D-glucosidic linkages in polysaccharides to remove successive maltose units from the non-reducing ends of the chains (EC 3.2.1.2).

Bio-based product: A product wholly or partly derived from materials of biological origin, excluding materials embedded in geological formations and/or fossilized (CEN definition).

Biocatalysis: Use of biological systems or their parts to catalyze chemical reactions.

Bioeconomy: Encompasses the production of renewable biological resources and the conversion of these resources, residues, by-products, and side streams into value-added products, such as food, feed, bio-based products, services, and bioenergy.

Biomass: Material of biological origin, excluding the material embedded in geologic formations and/or fossilized.

Biorefinery: Facility that integrates biomass conversion processes and equipment to produce energy (fuels, power, and heat) and chemicals and materials from biomass.

β-Limit dextrin: The remaining polymer produced by enzymatic hydrolysis of amylopectin with β-amylase, which cannot hydrolyze the α-1,6 bonds at branch points.

Blocklet: Structure mainly formed by crystalline and amorphous lamellae within the growth rings; blocklets are more or less spherical and have a size in the order of 100 nm. Blocklets constitute a unit of the growth rings.

Building block (amylopectin): The smallest, branched units found inside clusters of amylopectins. Amylopectin branch points are present in amorphous lamellae of starch granules and organized into densely branched areas, referred to as building blocks. One single amylopectin cluster contains several building blocks.

C

Catabolism: General term for the enzyme-catalyzed reactions in a cell by which complex molecules are degraded to simpler ones with the release of energy.

Chains in amylopectin: A-Chain is not branched; B-chain is branched; C-chain contains the reducing end.

Chair conformation: The six-membered ring of monosaccharides exists in two isomeric chair conformations, which are specified as $^1C_4$ and $^4C_1$, respectively, where the letter C stands for 'chair' and the numbers indicate the carbon atoms located above or below the reference plane of the chair, made up by C2, C3, and C5.

Chemiosmosis: The movement of ions across a semi-permeable membrane, down their electrochemical gradient.

Chloroplast: An organelle found in plant cells and eukaryotic algae that conducts photosynthesis.

Chloroplastic: Of or pertaining to a chloroplast.

Chyme: The pulpy acidic fluid that passes from the stomach to the small intestine, consisting of gastric juices and partly digested food.

Citric acid cycle or Krebs cycle: A series of chemical reactions used by all aerobic organisms to release stored energy through the oxidation of acetyl CoA into carbon dioxide and chemical energy in the form of ATP.

Coenzyme A: A small molecule used in the enzymatic transfer of acyl groups in the cell.

Coiled coil: Structural motif in proteins in which two to seven α-helices are coiled together like the strands of a rope (dimers and trimers are the most common types). Coiled-coil proteins contain a characteristic seven-amino acid repeat.

Colonic: Relating to or affecting the colon.

Colonocyte: An endothelial cell of the large intestine.

Cytoplasm: All of the material within a cell, enclosed by the cell membrane, except for the cell nucleus.

Cytosol: The liquid found inside cells around the organelles.

D

Dextrins: A group of low-molecular-weight carbohydrates produced by the hydrolysis of starch or glycogen. They are mixtures of polymers of D-glucose units linked by α-1,4- or α-1,6-glycosidic bonds.

Digestion: The breakdown of large insoluble food molecules into small water-soluble food molecules.

Dikinases: A category of enzymes that catalyze the chemical reaction:

$$ATP + X + Y \rightarrow AMP + XP + YP$$

Dikinases (E.C. 2.7.9) are all phosphotransferases.

Disproportionating enzyme (D-enzyme or DPE1): The DPE1 enzyme is a plastidial α-1,4-glucanotransferase; it catalyzes the transfer of α-1,4 linked oligosaccharides from the end of one glucan chain to the end of another. The shortest substrate on which it can act is maltotriose, but it can also act on a very large glucan molecule.

Dried distillers grains with solubles (DDGS): A common type of distillers grains, which are a cereal by-product of the distillation process.

Dual-specificity phosphatase: A family of phosphatase that dephosphorylates phospho-tyrosine, or phospho-serine/threonine of proteins, as well as more diverse substrates such as lipids, nucleic acids, or glucans.

Duodenum: The beginning portion of the small intestine.

E

Endosperm: The tissue produced inside the seeds of most of the flowering plants following fertilization. It surrounds the embryo and provides nutrition in the form of starch, though it can also contain oils and protein.

Eukaryote: An organism whose cells have a nucleus enclosed within membranes.

F

Fermentation: A metabolic process that consumes carbohydrates in the absence of oxygen through the action of enzymes produced by microorganisms, like yeast and bacteria. **Fermentation in food processing** is the process of converting carbohydrates to alcohol or organic acids using microorganisms under anaerobic conditions. **Ethanol fermentation**, also called alcoholic fermentation, is a biological process that converts sugars such as glucose, fructose, and sucrose into cellular energy, producing ethanol and carbon dioxide as by-products. Zymase is an enzyme complex in yeasts that catalyzes the ethanol fermentation of sugars. The overall chemical reaction for alcoholic fermentation is:

$$C_6H_{12}O_6 \rightarrow 2C_2H_5OH + 2CO_2$$

First-generation biorefinery: Biorefinery that uses edible biomass as feedstock.

Forward genetics: The molecular genetics approach of determining the genetic basis responsible for a phenotype.

G

γ-Amylase (or amyloglucosidase; exo-1,4-α-glucosidase; glucoamylase, EC 3.2.1.3): Cleaves α-1,6-glycosidic linkages, as well as the last α-1,4-glysosidic linkages at the non-reducing end of amylose and amylopectin, yielding glucose.

Gelatinization: Starch gelatinization is a process of breaking down the intermolecular bonds of starch molecules in the presence of water and heat, allowing the hydrogen bonding sites to engage more water. This irreversibly dissolves the starch granule in water.

Germination: The process by which an organism grows from a seed or similar structure.

α-Glucan, water dikinase (GWD): A plastidic enzyme (EC 2.7.9.4) that catalyzes the chemical reaction:

ATP + α-glucan + H2O $\rightleftharpoons$ AMP + phospho-α-glucan + phosphate

4-α-Glucanotransferase: An enzyme that transfers a segment of a 1,4-α-D-glucan to a new position in an acceptor carbohydrate, which may be glucose or a 1,4-α-D-glucan.

Glucopyranose: A simple sugar that contains a six-membered ring consisting of five carbon atoms and one oxygen atom.

Glucosyl: Radical or substituent structure obtained by removing the hemiacetal hydroxyl group from the cyclic form of glucose.

Glycogen: A highly-branched polysaccharide of glucose that serves as energy storage in humans, animals, fungi, and bacteria. Glycogen has a molecular weight between 10^6 and 10^{7Da} (approximately 30,000 glucose units). Most of the glucose units are linked by α-1,4-glycosidic bonds, approximately 1 in 12 glucose residues also makes an α-1,6-glycosidic bond with a second glucose molecule which results in the creation of a branch. Glycogen is made up of only one molecule, whereas **starch** is made up of two. Glycogen has a branched structure while starch has both chain and branched components.

Glycogenin: An enzyme involved in converting glucose to glycogen. It acts as a primer, by polymerizing the first few glucose molecules, after which other enzymes take over. It catalyzes the reaction:

$$UDP - \alpha\text{-glucose} + \text{glycogenin} \rightleftharpoons UDP + \alpha\text{-glucosylglycogenin}$$

Glycolysis: Ubiquitous metabolic pathway in the cytosol in which sugars are incompletely degraded with the production of ATP.

Glycosidic linkage: A type of covalent bond that joins a carbohydrate molecule to another group, which may or may not be another carbohydrate. A glycosidic bond is formed between the hemiacetal group of a saccharide and the hydroxyl group of some compounds such as an alcohol. A substance containing a glycosidic bond is a glycoside. The 1,4 glycosidic bond (1,6 bond) is formed between the C1 of one monosaccharide and C4 (C6) of another monosaccharide. There are two types of glycosidic bonds: α and β.

Glycosyl: A radical or substituent structure obtained by removing the hemiacetal hydroxyl group from the cyclic form of a monosaccharide.

Glycosyltransferase (GT): An enzyme that catalyzes the transfer of a monosaccharide residue from an activated donor substrate to an acceptor molecule, forming glycosidic bonds.

Granule: A small piece like a grain of something.

Greek-key barrel: A common domain structure in proteins is the Greek key β-barrel, a type of antiparallel β-barrel, where two Greek key motifs fold together to form an eight-stranded antiparallel β-barrel.

Greek-key motif: Consists of four adjacent antiparallel strands and their linking loops. It consists of three antiparallel strands connected by hairpins, while the fourth is adjacent to the first and linked to the third by a longer loop.

Growth ring: Alternating zones of semi-crystalline and amorphous material within starch granules.

GT-B fold: Fold exhibited by glycosyltransferases (GT). The GT-A fold consists of two dissimilar domains, one involved in nucleotide binding and the other binding the acceptor. The GT-B fold consists of two similar Rossmann fold subdomains. The Rossmann fold is a structural motif found in proteins that bind nucleotides. It is composed of alternating β-strands and α-helical segments.

H

Herb: A plant with savory or aromatic properties that is used for flavoring and garnishing food, medicinal purposes, or for fragrances.

Heterotroph: An organism that cannot produce its own food, relying instead on the intake of nutrition from other sources of organic carbon, mainly plant or animal matter.

Homeostasis: In biology, homeostasis is the state of steady internal physical and chemical conditions maintained by living systems.

Hylum: Point of origin of the ring structure.

I

Isoform: A member of a set of highly similar proteins that originate from a single gene or gene family and are the result of genetic differences.

L

Lamella (starch): The whole starch granule consists of stacks of semi-crystalline regions that are separated by amorphous growth rings. In each partially crystalline ring, there are alternating crystalline lamellae and amorphous lamellae. The crystalline lamellae are composed of double helices formed from outer chains of amylopectin, whereas the amorphous lamellae are made up of glucose units near branch points of the amylopectin molecules.

Legume: A plant in the family Fabaceae (or Leguminosae) or the *seed* of such a plant (also called pulse). Grain legumes include beans, lentils, lupins, peas, and peanuts.

Limit dextrin: When a branched polysaccharide such as glycogen or amylopectin is hydrolyzed enzymatically, glucose units are removed until a branch point is reached. The hydrolysis then stops, leaving what is termed a limit dextrin; further hydrolysis requires a different enzyme.

LSF2 (Like Sex Four2): Phosphoglucan phosphatase, chloroplastic.

Lysophospholipid: Small bioactive lipid molecule characterized by a single carbon chain and a polar head group.

M

Maltase: An enzyme that catalyzes the hydrolysis of the disaccharide maltose to glucose.

Maltodextrin: A polysaccharide consisting of D-glucose units connected in chains of variable length. The glucose units are linked with α-1,4-glycosidic bonds. Maltodextrin is typically composed of a mixture of chains that vary from 3 to 17 glucose units long.

Maltogenic amylase or glucan α-1,4-maltohydrolase: An amylolytic enzyme that catalyzes the exohydrolysis of α-1,4-glucosidic bonds in amylose, amylopectin, and related polymers. Maltose units are successively removed from the non-reducing end of the polymer chain until the molecule is completely degraded or, in the case of amylopectin, a branch point is reached.

Maltose: A disaccharide formed from two units of glucose joined with an α-1,4-glycosidic bond.

Maltotetraose: A tetrasaccharide consisting of four glucose units linked with α-1,4-glycosidic bonds.

Maltotriose: A trisaccharide consisting of three glucose molecules linked with α-1,4-glycosidic bonds.

Mash: A soft mass made by crushing a substance into a pulp, sometimes with the addition of liquid.

Microbiota: Microorganisms of a particular environment.

N

Nucleoside: Molecule containing a nitrogenous base bound to a five-carbon sugar (either ribose or deoxyribose).

Nucleotides: Molecules consisting of a nucleoside and a phosphate group. They are the building blocks of DNA and RNA.

Null mutation: A type of mutation in which the altered gene product lacks the molecular function of the wild-type gene.

O

Organelle: In cell biology, an organelle is a specialized subunit within a cell that has a specific function.

Ortholog: One of two or more genes in different species that are similar to each other because they originated from a common ancestor; ortholog retains the same function in the course of evolution.

Oxidative phosphorylation: Process in bacteria and mitochondria in which ATP formation is driven by the transfer of electrons from food molecules to molecular oxygen.

P

Pancreas: A large gland behind the stomach, which secretes digestive enzymes into the duodenum.

Penicillium oxalicum: An anamorph species of the genus of *Penicillium*.

Phosphatase: Phosphatase catalyzes the hydrolysis of a phosphomonoester, removing a phosphate group from the substrate. The general reaction catalyzed by a phosphatase enzyme is:

Phosphoglucan, water dikinase: Enzyme (EC 2.7.89.5) that catalyzes the chemical reaction: $ATP + phospho-\alpha-glucan + H_2O$ $AMP + O-phospho-phospho-\alpha-glucan + phosphate$

Phosphorylase: Enzyme, EC 2.4.1.1, that catalyzes the addition of a phosphate group from an inorganic phosphate H-OP to an acceptor (A-B + H-OPA-OP + HB). PDB 1Z8D. It catalyzes the cleavage of a glycosidic bond through substitution with phosphate.

Phosphorylase-limit dextrin: The polysaccharide produced by the action of starch phosphorylase on the branched polysaccharide amylopectin. These dextrins have three D-glucosyl residues on the external B-chains and four residues on the A-chains.

Photoassimilate: One of many biological compounds formed by assimilation using light-dependent reactions.**Phytoglycogen:** A type of glycogen extracted from plants. It is a highly branched, water-soluble polysaccharide derived from glucose.

Photosynthate: Any substance synthesized in photosynthesis, especially a sugar.

Plastid: A double-membrane organelle found in the cells of plants, algae, and some eukaryotic organisms. Plastids that contain chlorophyll can carry out photosynthesis and are called chloroplasts. Plastids in non-photosynthetic tissues are the site of fatty acid, starch, and amino acid syntheses.

Post-translational modification of proteins: Refers to the chemical changes proteins may undergo after translation, i.e. after protein synthesis. Such modifications come in a wide variety of types and are mostly catalyzed by enzymes that recognize specific target sequences in specific proteins.

Prebiotics: Prebiotics are compounds in food that induce the growth or activity of beneficial microorganisms. The most common example is in the gastrointestinal tract, where prebiotics can alter the composition of organisms in the gut microbiome. Dietary prebiotics are typically non-digestible fiber compounds that pass undigested through the upper part of the gastrointestinal tract and stimulate the growth or activity of advantageous bacteria that colonize the large bowel by acting as a substrate for them.

Probiotics: Live microorganisms intended to provide health benefits when consumed, generally by improving or restoring the gut flora; they are naturally created by fermentation in foods like yogurt.

Prokaryotes: A unicellular organism that lacks a membrane-bound nucleus, mitochondria, or any other membrane-bound organelle. Prokaryotes are divided into two domains, archaea and bacteria.

Protein tyrosine phosphatases (PTP): A group of enzymes that remove phosphate groups from phosphorylated tyrosine residues on proteins.

Pullulan: A polysaccharide consisting of maltotriose units, also known as α-1,4-;α-1,6-glucan'. Three glucose units in maltotriose are connected by an α-1,4-linkage, whereas the maltotriose units are connected by an α-1,6-linkage.

R

Regrowth: The growing back of hair, plants, etc.

Repressor: A DNA- or RNA-binding protein that inhibits the expression of one or more genes by binding to the operator or associated silencers.

Resistant starch: Starch, including its degradation products, that escapes from digestion in the small intestine of healthy individuals.

Retrogradation (starch): Gelatinized starch, when cooled for a long enough period (hours or days), will thicken and rearrange itself to a more crystalline structure following a process called retrogradation.

Reverse genetics: Reverse genetics is a method in molecular genetics that is used to help understand the function of a gene by analyzing the phenotypic effects of specific nucleic acid sequences after being genetically engineered.

Rhizome: Modified subterranean plant stem that sends out roots and shoots from its nodes.

Rossmann fold: A structural motif found in proteins that bind nucleotides.

S

Saccharification: The hydrolysis of polysaccharides to form simple sugars.

Sago: Either edible starch, which is obtained from a palm and is a staple food in parts of the tropics, or the palm from which most sago is obtained, growing in freshwater swamps in SE Asia.

Second-generation biorefinery: Biorefinery that uses non-edible feedstock as biomass.

Seed: An embryonic plant enclosed in a protective outer covering.

SEX4 (Starch Excess4): Dual specificity phosphatase family protein. SEX4 contains an amino-terminal chloroplast targeting peptide (cTP), followed by a dual-specificity phosphatase (DSP) domain, and a carbohydrate-binding module (CBM).

Sorghum: A genus of flowering plants in the grass family Poaceae. It is a cereal and is a major source of grain and stock feed.

Source organ: Photosynthetically active net exporters of photoassimilates, represented mainly by mature leaves, as opposed to sink organs that are photosynthetically inactive and referred to as net importers of fixed carbon.

Sprout: To produce leaves, hair, or other newly developing parts or (of leaves, hair, and other developing parts) to begin to grow.

Storage organ: A part of a plant specifically modified for storage of energy (generally in the form of carbohydrates) or water. Storage organs often grow underground, where they are better protected from attack by herbivores.

Stroma: A colorless fluid surrounding the grana within the chloroplast.

Sucrose: A non-reducing disaccharide composed of glucose and fructose residues.

T

TILLING: Targeting-induced local lesions in genomes; a method in molecular biology that allows directed identification of mutations in a specific gene.

Transcription: The first step in gene expression. It involves copying a gene's DNA sequence to make an RNA molecule.

Transcriptional regulation: In molecular biology and genetics, transcriptional regulation is how a cell regulates the conversion of DNA to RNA (transcription), thereby orchestrating gene activity.

Transgenesis: The process of introducing foreign genetic material, such as DNA or RNA, into host cells.

Transit peptides: N-terminal extensions that facilitate the targeting and translocation of cytosolically synthesized precursors into plastids via a post-translational mechanism.

Translation: Synthesis of a protein from an mRNA template.

Translocation: The movement of materials from leaves to other tissues throughout the plant.

V

Vacuole: Membrane-bound organelle which is present in all plant and fungal cells and some protist, animal, and bacterial cells.

V-Polymorph: A crystalline complex formed by single amylose helices with amphiphilic or hydrophobic ligands.

W

Waxy: Amylose-free.

Y

Yeast: Eukaryotic, single-celled microorganism classified as member of the fungus kingdom.

Z

Zymomonas mobilis: A Gram-negative rod-shaped bacterium. It has notable bioethanol-producing capabilities, which surpass yeast in some aspects.

Index

Note: **Bold** page numbers refer to tables and *italic* page numbers refer to figures